THE SOUTH CHINA SEA WAR

THE SOUTH CHINA SEA WAR

THE WAR NO ONE SHOULD WANT

FREDERICK M. BANKER

M·P·P
MissionPointPress.com

Mission Point Press

Published by Mission Point Press
www.MissionPointPress.com

Hardcover ISBN 13: 978-1-968761-56-1
Paperback ISBN 13: 978-1-968761-57-8

LCCN: 2026907340

Printed in the United States of America

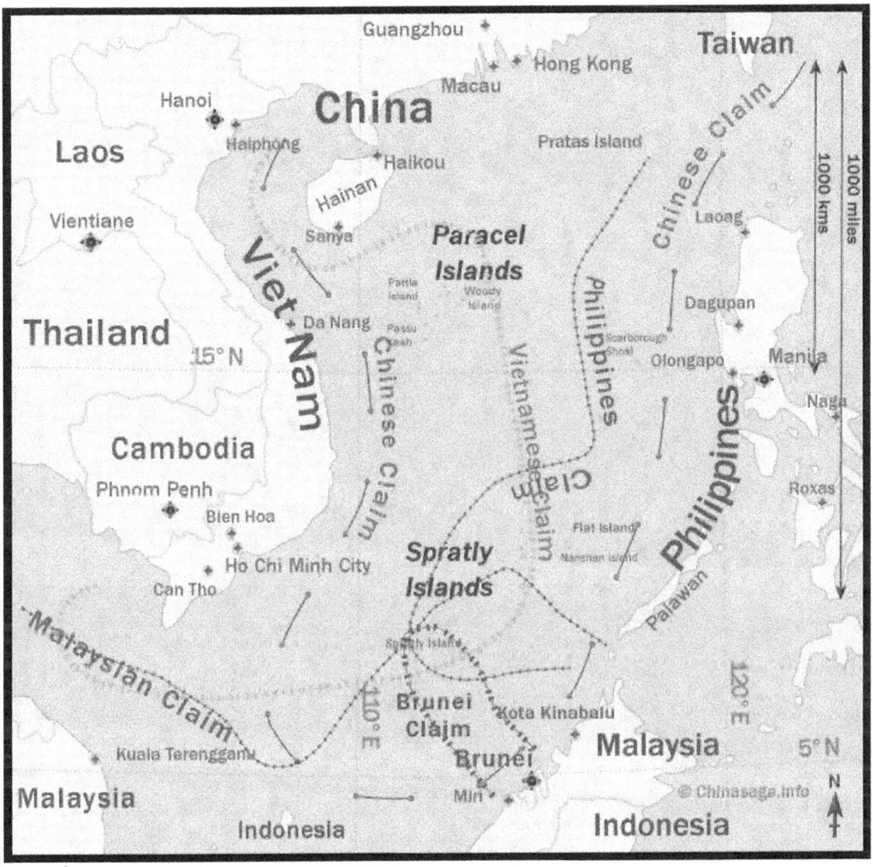

Source: https://www.chinasage.info/south-china-sea.htm

For my family, friends, and professors
who supported me throughout my life

"The century of humiliation at the hands
of the Western powers is a
period etched in acid on the pages
of Chinese student textbooks today."

—Robert D. Kaplan, *Asia's Cauldron:*
The South China Sea and the End of a Stable Pacific

Preface

Ever since I was a child, I have been fascinated with history. As I was getting into the subject, I noticed the rise in China, with everything I buy being made in China, and China hosting the 2008 Beijing Olympics. No matter how much time passed, the rise of China remained as a regular topic among history and political commentators on the news and on YouTube. China has been in the back of my mind as its actions concerning internal and external spheres would be a talking point for a decade to come.

While a student at Ferris State University, I was wondering what the topic for my book would be. My father suggested that I should cover a recent topic. Given my fascination with China and that nation's actions in the South China Sea, I saw a good opportunity for a book about the situation in the region. Having two grandfathers serve in the army and seeing how much damage war has caused in Iraq, Afghanistan, and Ukraine, I wrote this book to help people understand that we do not need a war in the South China Sea.

Throughout writing this book, I found it imperative to bring up issues that are currently causing distress in the region and to consider how they would be exacerbated if a major war occurred in the South China Sea. It was also during this time that I was supported by my family, friends, college mates, and college professors, who have encouraged me to go further into my research.

For the past 10 years, tensions have been rising in the South China Sea, putting the United States of America and the People's Republic of China at a standoff regarding the future of the region. Other countries have been forced to take sides as to who should be the dominant force in the South China Sea. With the plethora of nations in play and the news networks with their own agendas, it is difficult for people to keep track of what is going on in the region and who is doing what. The book provides a definitive description of the South China Sea, why the region is important to the world, and how peace and security can be maintained in the region.

Frederick Banker

Table of Contents

Introduction: The Potential of War and the Potential to Avoid It

The "War" in the title of this book has so far been contested primarily through trade policies, political bluster, and minor confrontations on disputed islands. Yet, the prospect of a catastrophic military conflict hangs heavily over one of the globe's most prosperous and highly populated regions.

The People's Republic of China, Vietnam, The Philippines, and other nations compete to control portions of the South China Sea, while the island of Taiwan seeks to defend its de facto independence from China. Meanwhile, outside powers including the United States, Japan, Australia and India have employed military alliances and economic incentives to exert influence in the region.

This book is intended to examine the current situation and to explore the prospects of peace or war in the South China Sea, where military and economic issues are overlain with long-standing cultural and historical grievances.

Since the Cold War (1945-1991), Southeast Asia has developed into one of the most economically prosperous regions in the world. By mining raw resources, attracting foreign investors, and building up trade ports, the nations around the South China Sea have enjoyed an era of relative peace that has lasted over 50 years. This gave the rest of the world a stable area for trade and give millions of people the food and resources needed to live. Most of those resources come from the sea trade routes. The nations of the South China Sea rely on a peaceful accord so that all nations and their people can live freely.

Such accord is threatened by the government of the People's Republic of China, which wants to control the entire region and its resources. China has claimed the entire South China Sea, overlapping the claims of the other nations in the region. The government has gone as far as assaulting the fishing vessels of the other nations in the region, cutting communication cords belonging to other countries, and overfishing in the waters of other nations. This has raised concerns about the future of the region, as the growing populations around the South China Sea will need all the food and resources they can get.

The United States has participated in South China Sea affairs since the late 1800s, and despite being located half a world away, is deeply concerned about China's actions in the region. It does not help that China is the only country in the world that appears able to match America's military and economy. In the Pacific region around the South China Sea, the United States has military bases in Japan and South Korea, and military relationships with India and Australia, which have had their own grievances with China. In return, China has strengthened relations with Russia, North Korea, and Pakistan to counter the nations that have partnered with the United States. The United States sees its investments in the South China Sea as a way to support its military partners in order to better patrol the region's international waters.

The one linchpin that brings the United States and China to a standoff is the island nation of Taiwan, which was a part of China until the Communist revolution that took place after World War II. The Taiwanese government is in a political limbo as most nations in the world recognize China as a singular nation and Taiwan as a part of the People's Republic of China. China wants to integrate Taiwan into the greater fold of China and put an end to what they see as a breakaway province.

This leaves Taiwan to rely on arms trade with the United States to keep up with the military on the Chinese mainland. Given the lack of support for integration in Taiwan, the Chinese government is preparing for a potential amphibious invasion of the island nation. The United States wants to protect Taiwan to contain the Chinese navy and dissuade China from aggression. Taiwan has also made strides to defend itself. Not only have the Taiwanese invested in modern military weapons and supplies, but the nation has also invested in chip manufacturing and now produces a majority of semiconductors used in modern-day electronics. This makes Taiwan indispensable on the world stage as more countries rely on semiconductors for everyday life.

With so many nations involved in the South China Sea, a major conflict in the region could spill over the rest of the world and be the powder keg for the third world war. Not only would this cause incalculable death and destruction, but it could cause a nuclear war as well. Even if the war was restricted to Southeast Asia, it would still destroy decades of progress in the region. For the sake of the people who have made the South China Sea one of the most prosperous parts of the world, a major war must not happen in the region.

Ever since the rise of China in the 1980s, the chances for a war between China and Taiwan have also risen. Historians and other scholars have produced decades worth of books and journal articles discussing the ever-looming threat of a war between Taiwan and China. The scenario of the continuation of the Chinese Civil War (1945-1949) has been treated as a potential reality by powers that rival China, such as America, Japan, India, and Australia. Everyone involved is waiting to see who will make the first move and how that will be countered.

The South China Sea has become the background of a standoff between some of the strongest nations in the world. The biggest players are the United States of America and the People's Republic of China. The former wants to keep the status quo of freedom of trade and navigation, where the latter wants to establish a hegemony and push America and other major powers out of the region. The major powers with influence in the South China Sea are Japan, India, Australia, South Korea, and Taiwan which are aligned with the United States; and Russia, Pakistan, and North Korea, which side with China. Neutral nations include Indonesia, the Philippines, Vietnam, Cambodia, Laos, Thailand, Brunei, Singapore, Malaysia, and Myanmar. The European nations of Great Britian, France, and Germany are interested in the region and are closely observing the standoff.

The tension between China and Taiwan is regularly discussed in the mainstream media. What is less reported, but also significant, is China's intrusion into the South China Sea and the islands in it. As China's wealth has grown throughout the 2000s, so has its aggression. Nations surrounding the South China Sea, including Vietnam, Brunei, Malaysia, Indonesia, the Philippines and Taiwan have claims to islands in the Sea. China, meanwhile, lays claim to most of the sea, including islands claimed by other nations. The wealth of resources in the South China Sea and the revenue from trade make the region an attractive target for China to establish a hegemony. This leads to a country that is breathing down the necks of its southern neighbors. The territorial disputes create the dangerous prospect that war between China, Taiwan and their respective allies, could expand into a regional struggle encapsulating the South China Sea and its many nations.

Going west of the South China Sea, China shares a border with India, one of their biggest rivals. This opens the potential that war with China could become a land war along with a sea and air war. However, the Chinese know about this potential front along China's southwest border and have taken steps to address it by strengthening ties with Pakistan, who next to China, is India's biggest rival. This could open a second front that India would have to fight. Since India, Pakistan and China are all armed with nuclear weapons, this multi-front scenario increases the possibility of war with China becoming nuclear.

At the northeastern border of China lie the two Koreas. For now, both North and South Korea are content with keeping with the ceasefire established after the Korean War in 1953. Despite this, China can disrupt this peace by leveraging North Korea's dependence on Chinese goods to go to war with either South Korea or America and Japan. South Korea's own reliance on Chinese goods, albeit at a lesser extent than North Korea, opens the probability that South Korea could be forced into neutrality by China. However, given South Korea's recent rise in military power and growing ties with America and its allies, a war with China could reignite the Korean War, opening up another front that would force both America and China to divert supplies and focus from Taiwan.

Further north are the far-eastern regions of Russia. While the Russians are currently dealing with their war with Ukraine, Russia is staying firmly in partnership with China. Russia also has disputed land in the Pacific. Both it and Japan have pursued some claim to the Kuril Islands north of Japan's Hokkaido region. Taken together with Russia's rivalry with the United States, this demonstrates one of several reasons the Russian government could get involved in the war on the side with China. Though it is debatable as to how much Russia will be able to impact the South China Sea War, its involvement will still mean the addition of another major nuclear power in the conflict.

Of course, it should not be forgotten that America is as much involved in the South China Sea as any other nation. Having ties with nations in the South China Sea, and bases in Japan and South Korea, America has invested heavily in relations, bases and territories in the Pacific for more than 100 years. The United States also has defensive treaties with Japan and the Philippines, which means that if they were attacked, the United States would come and fight on their side. Both countries have been threatened by China in the past. America has been China's biggest rival in the South China Sea, as China wants to kick America out of the region. It does not help matters that America is supplying Taiwan with military aid that would be used in the event of a war breaking out between Taiwan and China. America has defended the post-WW2 status quo for nearly 80 years and there is no sign that that will change anytime soon.

Outside the South China Sea is a continent that has much interest in the region as the nations in it: Europe. Especially the countries of Britain, France, and Germany, who have the same interest as America to curb Chinese aggression. Western Europe has helped Southeast Asian nations with weapons and other supplies to combat the threats and acts of aggression they see as saber-rattling from China. However, much like Russia, at the time of writing this, the European nations are preoccupied with the Ukrainian conflict, and their fears that Russia is becoming more desperate and more dangerous. While Europe would not be involved with the potential conflict, they could play a more supportive role in the said conflict.

This potential war can plunge not only an entire region into war, but the whole world. However, this does not mean that this war needs to happen. Though the possibility of war has been built up through 70 years of geopolitical events, conflict has so far been avoided because of the consequences that it would have on all parties involved. Not only would it disrupt trade in the South China Sea, but with all sides preparing for conflict there would be no easy winners. This can be shown by China's cautious nature when it comes to taking on Taiwan, waiting for the right time to strike. While it is important to investigate the military capabilities of these nations, it is relevant to showcase how these nations can reach a peaceful agreement.

To delve further into this situation, the book has been sectioned into 17 chapters, the first and last being the introduction and conclusion. Chapter 1 covers the historical events that got the South China Sea to this point. Chapters 2-11 look at countries that would have the most proactive roles in the South China Sea War. Chapter 12 goes into the formation of the Quadrilateral Security Dialogue (QSD or QUAD) and how the QUAD members cooperate in the South China Sea. Chapter 13 discusses the Association of Southeast Asian Nations (ASEAN): how this organization views the geopolitical situation in the region and how it might respond in the event of war. Chapter 14 describes the interests of European nations and how those nations' actions could affect the war. Chapter 15 considers two potential scenarios if war were to break out. Chapter 16 delves into the idea that war can be avoided in the South China Sea, and Chapter 17 summarizes the author's thoughts on the proceeding chapters.

Chapter 1:
1949-Now: How We Got to Where We Are Now

The first point of contention in the South China Sea is the China-Taiwan standoff. Dr. Lowell Dittmer wrote in his book that the history of China and Taiwan can be divided into four stages. The first stage (December 1949-October 1971) had China and Taiwan in a military standoff with each other. Due to its alliance with the US and being the only Chinese nation recognized by the United Nations, Taiwan was able to make connections with Thailand, the Philippines, and the former Republic of Vietnam (South Vietnam). This left China isolated and making connections with Communist countries like North Korea and North Vietnam.

In the second stage (October 1971-July 1987), China was admitted into the UN, which removed Taiwan's political leverage. In 1972, China and the United States agreed to a "One-China" policy, in which a country could choose to recognize either mainland China or Taiwan, but not both. This left Taiwan diplomatically stranded. However, Taiwan continued to send representatives to the 23 countries that recognized its legitimacy.

In stage three (1987-1997) of Sino-Taiwanese history, the martial law which Taiwan had implemented in case of an attack from China was lifted in July of 1987. In May 1991, the Taiwanese ended their mobilization effort, formally ending the civil war between themselves and the Chinese. 1991 would mark the beginning of slow efforts to formalize relations between the two nations. From March to December

1991, the Straits Exchange Foundation (SEF) was founded in Taipei and the Association for Relations Across the Taiwan Straits (ARATS) was founded in Beijing.

In the fourth stage (1997-Present), China continues to grow economically, and Taiwan grows at a slow rate; showing the rise of China as the world superpower and Taiwan's continued political isolation.[1] As China has risen into power, Taiwan has lost much of its international soft power.

Throughout all the stages, Taiwan's determination to remain independent from China has been a constant thorn in relations. As explained by Dr. Dittmer, as of 2016, nearly two-thirds of people in Taiwan considered themselves Taiwanese. Only 33.6 percent of Taiwanese people consider themselves Chinese.[2] It did not help matters that when a bill that would further cross-strait integration (further integration of Taiwan into the Chinese state) was proposed in 2014, it led to Taiwanese students occupying the Legislative Yuan between March and April of that year.

Taiwan's ruling party at the time, the Kuomintang (KMT) suffered losses in local elections in 2014 and the presidential and parliamentary elections in 2016, largely due to the party's support for cross-strait integration to grow the economy and its failure to stop anti-China sentiment from spreading throughout the country. This gave the Democratic Progressive Party (DPP) a majority in the parliament for the first time and led to the start of the Tsai Ing-wen administration.[3] Despite a desire for peace, Taiwan's insistence on independence has continued to negatively impact relations between China and Taiwan.

Bilateral relations deteriorated as China grew agitated at Taiwanese

---

1 Dittmer, Lowell (Professor Dittmer received his Ph.D. from The University of Chicago in 1971. His scholarly expertise is the study of contemporary China. He teaches courses on contemporary China, Northeast Asia, and the Pacific Rim.), *Taiwan and China Fitful Embrace*. Published by the University of California Press (luminosoa.org) Oakland, California, U.S.A. 2017, 249-250.

2 Dittmer. 23.

3 Dittmer. 210.

expressions of independence. Dr. Dittmer displayed this with Xi Jinping saying that both China and Taiwan "share the same destiny" and should "strengthen their pride in the Chinese nation as well as their shared goal of its rejuvenation." Xi also said that "though the mainland and Taiwan are yet to be reunified, they belong to one China and are inseparable parts of the country."

In September 2014, Xi added on to the Taiwan Question by calling for "peaceful reunification; One Country, Two Systems." He added, "No secessionist act will be tolerated. The path of 'Taiwan independence' is unfeasible." It does not help matters that most countries (including America) recognize Taiwan as a part of China.[4] However, it must be said that most countries do not support the conquest of Taiwan. The conflict between China's adamant reunification efforts and Taiwan's unwavering commitment to independence creates the main point of contention in the South China Sea.

China's relationship with Russia is another important geopolitical issue, given that both China and Russia are seen as rivals to the United States. Artyom Lukin, an associate professor at the Far Eastern Federal University School of Regional and International Studies at Vladivostok, Russia, notes the significance of the China-Russia Treaty of Good-Neighborliness and Friendly Cooperation, signed in 2001. The agreement's language could provide grounds for a military alliance as in Article 9: "When a situation arises in which one of the contracting parties deems that peace is being threatened and undermined or its security interests are involved or when it is confronted with the threat of aggression, the contracting parties shall immediately hold contacts and consultations in order to eliminate such threats."

Similar wording in treaties against common threats is common in Western treaties such as Article 5 of both the NATO alliance and the US-Japan Security Treaty in 1960.[5] This makes it a dangerous prospect

4 Dittmer. 243-244, 285.

5 Lukin, Artyom (Associate Professor at Far Eastern Federal University School of Regional and International Studies at Vladivostok, Russia), China's Rejection

to stand up to China and Russia, especially in their own backyard, as the wording is vague as to what could be considered a threat to peace. However, given China's inactivity to Russia's war on Ukraine, despite Russia claiming that Ukraine under its leader Volodymyr Zelensky was threatening them, it is safe to say that either Russia or China could pick and choose when to support each other.

Meanwhile the United States has had land in Southeast Asia since the late 1890s. David Scott, a doctor presenter on Indo-Pacific geopolitics at the NATO Defense College in Rome, Italy, writes on how America has been viewing the situation. He starts by claiming that since the 1950s, America has been referring to the Pacific Ocean as an "American Lake."[6] This is further backed up by how Americans have focused on the Pacific Ocean since the 1950s. The United States has been eyeing China as a potential military threat for the past 20 years. Scott mentions that in 2006, the Pentagon Quadrennial Defense Review stated that "Of the major and emerging powers, China has the greatest potential to compete militarily with the United States and field disruptive military technologies that could over time offset traditional US military advantages absent US counter strategy."

India has also risen with economic policy. Former Secretary of State Rex Tillerson said in 2017: "China, while rising alongside India, has done so less responsibly, at times undermining the international, rules-based order … It makes perfect sense that the United States—at this time—should seek to build on the strong foundation of our years of cooperation with India. It is indeed time to double down on a democratic partner that is still rising—and rising responsibly—for

---

of the South China Sea Arbitration: "The Russia–China entente and its future." *International Politics* Vol. 58 363-380. Published by the Moynihan Institute for Global Affairs, Syracuse University. New York, U.S.A. June 13, 2020.

6 Scott, David (A doctor presenter on Indo-Pacific geopolitics at the NATO Defense College in Rome and the Baltic Defence College in Tartu, Estonia), "US Indo-Pacific Strategy: The Indo-Pacific in US Strategy: Responding to Power Shifts - Rising Powers in Global Governance" (kureselcalismalar.com), *Rising Powers Quarterly* Volume 3, Issue 2, 2018, 19-43, Published by Marmara University. Istanbul, Turkey. August 2018. 19.

the next 100 years … The Indo-Pacific in particular—needs the United States and India to have a strong partnership."

The United States has sought the help of other major powers in the Indo-Pacific to maintain the freedom of navigation as well. Former Secretary of State John Kerry explained that "the United States is already providing leadership on maritime security … in association with close friends and allies across the [Indo-Pacific] region, including India, Australia, Indonesia, and Japan."[7] For more than 20 years, America has paid attention to the rise of China and how to curb Chinese aggression. Scott discussed how America has developed ties with Vietnam, Singapore, and Indonesia. America also has conducted exercises with Japan.[8] America's presence in the South China Sea and the Indo-Pacific region makes China wary of attacking Taiwan and reinforces China's "with haste" attitude toward militarization.

A recent point of tension between China and the Association of Southeast Asian Nations (ASEAN), is the arbitration of the Spratly Islands and the Scarborough Shoal of 2013, both of which are west of the Philippines. Stefan Talmon, a professor of public international law at the University of Oxford in Oxford, England, and Bing Jia, a professor of international law at the Tsinghua Law School at Beijing, China, go further into the arbitration. In the book, Dr. Talmon and Professor Jia explained that on January 22, 2013, the Republic of the Philippines instituted arbitral proceedings against China under the United Nations Convention on the Law of the Sea (UNCLOS). The Philippines were contesting China's nine-dash line, which was the claim that China owned most of the South China Sea and the islands in the sea. This included the Spratly Islands and Scarborough Shoal, to which the Philippines has its own competing claims. China retaliated against the arbitral proceedings, noting that in 2002, the ASEAN-China DOC said that to resolve the territorial and jurisdictional disputes, a member state of ASEAN and China must commence

7 Scott. 21, 25, 27.
8 Scott. 23-24.

friendly and peaceful negations with each other, and that the claims of sovereignty are not the concern of the United Nations Convention on the Law of the Sea (UNCLOS).[9] This shows how smaller nations in the South China Sea stand up to China, while not having the military of America, Japan, or India.

This action marked 2013 as the turning point of how other nations in the Pacific looked at China. Wenjuan Nie, the deputy director of the Institute of International Relations at China Foreign Affairs University in Beijing, goes into further detail on the arbitration and its effect on relations between China and the nations of the South China Sea. Nie brings into focus how most nations in the South China Sea sided with the Philippines in the arbitration process. This included nations like Vietnam, Brunei, Malaysia, Indonesia, Japan, and America. Such actions did strain relations between China and the other South China Sea nations. Former Vietnamese President Truong Tan Sang stated his opposing opinion on China nine-dash-line claim, while visiting Washington, D.C. Japan, Singapore, and Indonesia also refuse to accept China's claims on the South China Sea.[10] The arbitration helped to unify the minor powers in the South China Sea in their efforts to curb Chinese aggression.

What caused a bigger blow to Chinese claims was the result of this arbitration. Professor Houlden, Scott N. Romaniuk, and Nong Hong, executive director of the Institute for China-America Studies, provide

---

9 Talmon, Stefan (Dr. Stefan Talmon LL.M. M.A. holds the Chair of Public Law, International Law and European Law. Prior to his appointment in October 2011, he was Lecturer, Reader and most recently Professor of Public International Law at the University of Oxford and Fellow of St. Anne's College, Oxford (2003-2011).), Jia, Bing (Professor of International Law Tsinghua Law School Beijing, China), *The South China Sea Arbitration: A Chinese Perspective*, Published by Hart Publishing 2014. 1-11.

10 Nie, Wenjuan (Deputy Director of Institute of International Relations, China Foreign Affairs University Beijing, China), One Belt One Road Or the South China Sea:Xi Jinping's Foreign Policy Dilemma: One Belt, One Road or the South China Sea? on JSTOR, *Contemporary Southeast Asia*, Vol. 38, No. 3 (December 2016). Published by ISEAS - Yusof Ishak Institute. Heng Mui Keng Terrace, Singapore. December 2016. 424-433.

further detail into what happened to the arbitration. After 3 years, UNCLOS sided with the Philippines on the Arbitration on July 12, 2016. This was followed up by America, Japan, and Australia issuing a joint statement calling "for the rule of law" and advising "China and the Philippines to abide by the Arbitral Tribunal's Award of July 12, 2016, in the Philippines-China arbitration, which is final and legally binding on both parties." India had their own statement, where they "urge[d] all parties to show utmost respect for the UNCLOS which [had] establish [ed] the international legal order of the seas and oceans."[11] The move by these major powers in the South China Sea to side with the Philippines shows not only who they side with in the legislative sense, but a geopolitical sense.

Another potential military flashpoint in the South China Sea is the Malacca Strait, a narrow waterway between Malay Peninsula and the Island of Sumatra. The Strait, which serves as the main shipping route between the Indian and Pacific Oceans, could become an additional theater of war, due to China's activities. Jiacheng Li, an associate professor at Liaoning University's School of International Studies in Liaoning, China, explains the Malacca Strait's integral role in China's economy. According to Li, China's three largest oil suppliers, Angola, Iran, and Saudi Arabia ship their product through the Malacca Strait. That makes trade through the Strait and going to China vulnerable in a time of war. In 2003, then Chinese President Hu Jintao, called this predicament the "Malacca Dilemma" at the Conference on Economic

---

11 Gordon, Houlden (Professor Gordon Houlden is Director Emeritus of the China Institute, Professor of Political Science and Adjunct Professor of the Alberta School of Business at the University of Alberta.), Romaniuk, Scott N. (Dr Scott N. Romaniuk is a Newton International Fellow of the British Academy at the University of South Wales' Faculty of Life Sciences and Education, and a Visiting Fellow at the university's International Centre for Policing and Security), Hong Nong (Executive Director of the Institute for China-America Studies and a Research Fellow at both the National Institute for South China Sea Studies, China, and the University of Alberta, Canada), *Security, Strategy, And Military Dynamics in the South China Sea: Cross-National Perspectives*. Published by Bristol University Press. Bristol, Britain. 2021. 297-298.

Work of the Communist Party of China. This "Dilemma" is not helped by the naval buildup of India, nor the United States's naval presence in the South China Sea. America has gone further by acknowledging the Malacca Strait as a point of weakness for China in a time of war, with a researcher at National Defense University in Washington, D.C. suggesting that if there was a war with China, then America would implement "sea control," blockade the strait, and send 13-15 Marine Corps units to intercept 800 ships taking goods to Chinese ports. This scenario would be intended to force China to end its belligerent behavior and come to the negotiation table.[12] The Malacca Dilemma is another reason that China is deferring military action until they are fully prepared for not only a war between Taiwan and their partners, but also for a multi-front war that could include the Strait of Malacca.

A non-territorial matter that has strained relations between America and China is America's use of "Freedom of Navigation" (FON) operations. Lynn Kuok described FON operations and explained why China does not like America's FON operations. According to Kuok, FON operations represent an effort to maintain the world maritime order that began in the 1970s as a response to newly independent nations claiming coastal islands. FON operations led by the US Department of Defense have a three-pronged approach: operational assertions by military units; diplomatic protests over excessive claims; and Department of State or Defense meetings with representatives of other states to promote maritime stability and consistency with international law.

China came into the equation in February 1992, when China passed its Law on the Territorial Sea and the Contiguous Zone, which requires warships to ask permission for innocent passage. This was protested by America, which relies on military ships for the FON operations. It

---

12 Li, Jiacheng (an associate professor at Liaoning University's School of International Studies in Liaoning, China). "China's Indian Ocean Strategy: Developing China's Indian Ocean Strategy: Rationale and Prospects." *China Quarterly of International Strategic Studies* Vol. 03, No. 04, (2017). Published by World Scientific Publishing Co Pte Ltd. Singapore. 483-484.

did not help matters when in March of 2009, Chinese fishing vessels harassed the survey ship USNS *Impeccable*. Though China considers FON operations as illegal, this did not stop America from conducting these missions near China's artificial islands (which were built to strengthen Chinese claims in the South China Sea) from October 2015 to May 2016. The disagreement between China and America on these issues was explained as follows by a Chinese diplomat: "The difference between China and the United States is not a difference between freedom of navigation. We are very clear that we support freedom of navigation in accordance with international law, not only in the South China Sea, but also other parts of the world. The difference is that we think the United States takes freedom of navigation as an excuse. What their ships and planes have been doing is actually a close surveillance of China's coastlines and islands and reefs."[13] The unwillingness of both China and America to negotiate on the issue of FON operations brings more fuel to the fire in the possibility of a South China Sea War.

Though disputes over Chinese claims of the South China Sea have flared up recently, the claims have been an issue between China and surrounding nations for hundreds of years. Leishangthem Singh asserts that the root of China's claims date all the way back to the first Chinese Imperial Dynasty, the Han Dynasty, that discovered the South China Sea Islands. When European powers were surveying the Spratly Islands in the late 1800s, China protested. The claims would outlive the Chinese Imperial Dynasties, as the Republic of China (now Taiwan), claimed the Spratly and Paracel Islands as a part of the Guangdong Province in 1947. After Mao Zedong's forces prevailed in the Chinese Civil War, creating the People's Republic of China, they still insisted that the South China Sea Islands were theirs.

---

13 Kuok, Lynn (fellow at Brookings Institution's Center for East Asia Policy Studies, a visiting fellow at the Harvard Law School, and a senior visiting fellow at the Centre for International Law.), China Undermines Rule of Law in South China Sea:How China's actions in the South China Sea undermine the rule of law (brookings. edu), Global China Assessing China's Growing Role in The World, November 2019, Published by the Brookings Institution in Washington, D.C. III-10.

This went as far as a clash in 1988 between China and Vietnam over the Paracel Islands, which resulted in the deaths of dozens of Vietnamese sailors.[14] Maximilian Ernst brings into larger focus the clashes that China had with countries over the South China Sea. More specifically, Ernst spotlights how China has been at odds with the Philippines for decades. From 1994 to 1995, China occupied Mischief Reef in the Eastern Spratlys, which are in the Philippines' Exclusive Economic Zone (EEZ). This tension between China and the Philippines almost caused a war in 2012 over the Scarborough Shoal. That was avoided by the 2013 arbitration. These territorial clashes were tense and sometimes violent but were not controversial enough to cause a global crisis.[15] Singh's and Ernst's journal articles shed light on China's use of nationalism to bully its neighbors for their holdings in the South China Sea.

While China's actions and claims involving the South China Sea have been simmering issues for more than 20 years, this topic has been brought to the forefront as China has become more assertive recently. Khadeeja Ahmad, Sadia Rafique, and Khalid Butt, in their journal article on China's activities over the past two decades, explain that China's aggression ramped up in 2009, when China bought four Zubr Class Landing Craft supplied by Ukraine and Greece.[16] The purchases

---

14 Singh, Leishangthem. "China's Strategy in the South China Sea :China's Strategy On The South China Sea Disputes" on JSTOR. *World Affairs: The Journal of International Issues*, Vol. 21, No. 1 (SPRING JANUARY-MARCH 2017). Published by Kapur Surya Foundation. India. 84-85, 88.

15 Ernst, Maximilian (a researcher with the Centre for Security, Diplomacy, and Strategy [CSDS] at the Brussels School of Governance and a PhD candidate at the Vrije Universiteit Brussels. His research focuses on Asia-Pacific security, particularly Chinese foreign policy towards regional states.) "Control Through Cooperation? Assessing China's Economic and Military-Strategic Interests in the South China Sea.":jtms_9.2.1.pdf (vub.be), *Journal of Territorial and Maritime Studies*. Published by Vrije Universiteit Brussel. Brussels, Belgium. 2022. 9-11.

16 Ahmad, Khadeeja (M.Phil. Scholar at the Department of Political Science, Government College University, Lahore, Pakistan.), Rafique, Sadia (Assistant Professor at the Department of Political Science, Government College University, Lahore, Pakistan), Butt, Khalid (Director, Centre of Excellence China Studies,

appear to have been in preparation for an invasion of Taiwan. The authors cover the recent actions of China, as on January 15, 2021, when the Chinese military reacted to seeing the USS *Theodore Roosevelt* in the South China Sea by doing live-fire exercises in the region.[17] What these two acts show is that even while preparing to invade Taiwan, China continues to stake its claim to other regions in the South China Sea.

One point of Chinese expansionism is the buildup of its navy. James E. Fanell contends that China's aggression has coincided with the rise of its navy. Fanell covers how as of 2018, China's People's Liberation Army Navy (PLAN) was made up of 330 surface ships and 66 submarines. In comparison, on May 4th, 2018, the American Navy had 283 battle-force ships, including 211 surface vessels and 72 submarines. By 2030 it is estimated that the PLAN will total 550 ships and submarines. Fanell states that China has quickly met the standards of the United States Navy (USN). He asserts that the PLAN does not need to be better than the USN, but good enough to achieve more hits and to win any given battle. Ever since 2000, the Chinese have planned to build up their blue-water navy (that is, a navy that can operate beyond 1,500 nautical miles, or 1,726 miles of its coastline). The PLAN and the Chinese Coast Guard (CCG) have increased their military operations in China's first and second island chains since 2000. Fanell explained that during the 19th National Party Congress in 2017, President Xi Jinping emphasized the idea that Chinese rejuvenation involves the expansion of the capabilities

Chairperson, Department of Political Science and Dean, Faculty of Arts and Social Sciences at Government College University, Lahore, Pakistan.). "China's Policy towards South China Sea and its Geo-Political Impact (2009-2020). :War on Terror Partnership and Growing/Mounting/Increasing/Rising Militant Extremism in Pakistan" (researchgate.net), *A Research Journal of South Asian Studies* Vol. 36, No. 2, July – December 2021. Published by the Centre for South Asian Studies, University of the Punjab. Lahore, Pakistan. 2022. 426.

17 Ahmad, Khadeeja, Rafique, Sadia , Butt, Khalid. "China's Policy towards South China Sea and its Geo-Political Impact (2009-2020). :War on Terror Partnership and Growing/Mounting/Increasing/Rising Militant Extremism in Pakistan." 426.

of the PLAN. In 2015, China sent five ships near Alaska to threaten America while President Barack Obama was visiting the Arctic. In 2017, China went to Alaska again to do an apparent intelligence-gathering mission on the American testing of the Terminal High Altitude Area Defense (THAAD), a missile defense system aimed at taking out short- to medium-range missiles.[18] This shows how China has risen to America's level in a military sense, and why America will need help from other nations to contain China.

As complicated as the history with China and Taiwan is, the United States has a complicated relationship with Taiwan as well. This is uncovered by Lung-chu Chen in his book on US-China relations. Chen notes that President Jimmy Carter derecognized Taiwan and recognized the People's Republic of China in 1978. Not only did this mark the first time that America would cut off ties with an allied nation, but it would muddle the waters between Taiwan and America. However, America continued to support Taiwan. A year after Carter's decision, Congress passed the Taiwan Relations Act (TRA), which continued the sale of US arms to Taiwan.[19] While the United States continues to support Taiwan as of the writing of this book, the level of separation between Taiwan and America could hinder their ability to help each other in the scenario of a war with China. Supporting

---

18 Fanell, James E. (US Navy [Retired], served as a career naval intelligence officer whose positions included the senior intelligence officer for China at the Office of Naval Intelligence and the chief of intelligence for CTF-70, 7th Fleet, and the US Pacific Fleet. He retired from the Navy in 2015 and currently is a government fellow at the Geneva Centre for Security Policy in Geneva, Switzerland.) "China's Global Naval Strategy And Expanding Force Structure: Pathway to Hegemony." :CHINA'S GLOBAL NAVAL STRATEGY AND EXPANDING FORCE STRUCTURE on JSTOR, *Naval War College Review*, Vol. 72, No. 1 (Winter 2019), Published by the US Naval War College, Newport, Rhode Island, U.S.A. 2019. 13-14, 40-41.

19 Chen, Lung-chu (Scholar and Professor of Law at New York Law School, specializing in international law, human rights, the United Nations, and Taiwan). *The US-Taiwan-China Relationship in International Law and Policy.*: The US-Taiwan-China Relationship in International Law and Policy - Lung-chu Chen - Google Books. Published by Oxford University Press. New York, New York, U.S.A. 2016. XX.

Taiwan too much might push the Chinese to invade before they lose their chance.

China's determination to become the world's superpower, balanced by the United States's persistent efforts to keep that position, creates a geopolitical standoff that echoes an era when America stared at the Soviet Union until they blinked. While the Cold War was centered around Eastern Europe, the tension between America and China is mainly in the South China Sea. English author Geoff A. Dyer explains how both countries look at the South China Sea and why it is important that they remain there. Dyer brings up a quote from Chinese historian Yang Yong: "Ignoring the oceans is a historical error we [the Chinese] committed ... And now even in the future we will pay a price for this error."[20] This spotlights how the wars that China participated in from the mid-1800s to the early 1900s caused everlasting trauma to the Chinese, especially since they lacked the modern navy at the time to take on their enemies.

China intends to have a navy strong enough to take on all threats. This is supported by a 2005 PLA military doctrine statement called The Science of Military Strategy, which said that if the PLAN were restricted to the first island chain, then "the essential strategic space for China's rejuvenation will be lost." More recently, Chu Shulong, an academic at Tsinghua University, wrote that the grudge China has against the American navy has only grown stronger. Despite the advancements of the PLAN, Chinese officers still witness the superiority of the American navy on a daily basis, as they see United States ships within a few miles of China's major cities. Shulong added that "It is humiliating that another country can exercise so close to China's coasts, so close to the base in Hainan. That is the reason the navy wants to do something

---

20 Dyer, Geoff A. (Geoff Dyer is an English author. He has written a number of novels and non-fiction books, some of which have won literary awards.) *The Contest of the Century: The New Era of Competition with China—and How America Can Win.* :The Contest of the Century: The New Era of Competition with China—And How ... - Geoff A. Dyer - Google Books, Published by Vintage Books. New York, New York, U.S.A. 2014. 25.

to challenge the US." The combination of history and geopolitics leads to China feeling anxious about the American navy operating close to its coasts.

America's interest in the South China Sea began in the same era as China's Century of Humiliation (1842-1949), in which China endured losses in wars with European nations and gave out treaty ports to said nations. Dyer explains that America's first diplomatic move in the region was the signing of a treaty of commerce and amity between the United States and Siam (modern day Thailand) in 1833. Commodore Matthew Perry opened relations with Japan 20 years later.[21] The reason for these moves is that America was following the footsteps of Europe, which would start to normalize relations with China and other Asian nations after the era of colonialism between the 1500s and 1700s. Dyer notes that America became more invested in the South China Sea and the Indo-Pacific at large during the World Wars and the Cold War. Dyer does focus on an event, in 1996, when three Chinese warships made a goodwill port call in San Diego. The American strategists who looked into the ships saw this as a sign of China's primitive military and joked that the Chinese invasion of Taiwan would involve "a million-man swim." However, China's militarization since that time has made them more than a laughingstock, with estimates stating that China will have a higher defense budget than America in 2025.[22] While not having as long of a history with the South China Sea as China does, America's near 200-year investment in the region should not be scoffed at.

The Chinese-American standoff in the South China Sea has made other countries choose between the two big powers. Madeline McLaughlin showcases the dialogue between American and Chinese officials and how this rivalry affects the other nations in the South China Sea. McLaughlin reports on an exchange in July of 2020, in which former Secretary of State Mike Pompeo said: "The United States

21 Dyer. 32.
22 Dyer. 32-33, 44-45.

champions a free and open Indo-Pacific. Today we are strengthening US policy in a vital, contentious part of that region—the South China Sea. We are making clear: Beijing's claims to offshore resources across most of the South China Sea are completely unlawful, as is its campaign of bullying to control them." The Chinese Embassy responded with: "We advise the US side to earnestly honor its commitment of not taking sides on the issue of territorial sovereignty, respect regional countries' efforts for a peaceful and stable South China Sea and stop its attempts to disrupt and sabotage regional peace and stability."

While this was happening, during the summer of 2020, America, Japan, and Australia had a joint military exercise.[23] The regional tension created by the United States and China has made it hard for other countries in the South China Sea to remain neutral. McLaughlin explains that while the Philippines and America have a joint security pact, the Philippine president at the time, Rodrigo Duterte, wanted to work more with China than America. In July of 2020, Vietnam accepted a loan of six new patrol vessels from America.[24] The rivalry between the United States and China has its tremors felt throughout the region and will expand the war across the whole South China Sea.

The United States's reliance on FON operations (or FONOPS), naval exercises to protect trade routes, in the South China Sea is based on decades of traditional diplomacy. Retired American navy captain and professor at the Stockton Center for International Law in the US Naval War College at Newport, Rhode Island, Raul 'Pete' Pedrozo, discusses this tradition in a 2022 journal article. Pedrozo goes as far back as 1995 to show how America's approach to the South China Sea has not

---

23 McLaughlin, Madeline (the National Security Strategy intern at ASP. She graduated from the University of Utah in 2019 with a Fine Arts degree and a political science minor). "US Strategy in the South China Sea.":resrep26608.pdf (jstor.org). American Security Project (2020). Published by American Security Project in Washington, D.C., U.S.A. October 2020. 1-2, 4.

24 McLaughlin. 3-4.

changed in more than 20 years. He explains that the 1995 American South China Sea policy was based on four columns, or principles:

1) Oppose the use or threat of force to resolve competing claims.
2) Intensify diplomatic efforts to resolve the competing claims, considering the interests of all parties, and which contribute to peace and prosperity in the region.
3) Maintain freedom of navigation by all ships and aircraft in the South China Sea.
4) Take no position on the legal merits of the competing claims to sovereignty over the various features in the South China Sea, but view with seriousness with scrupulous concern any oceanic claim or restriction on maritime activity in the South China Sea that is inconsistent with international law, including UNCLOS.

Despite the age of this policy, all four of these columns become relevant to this day as China has broken all of them. From interfering in the oil and gas developments of Vietnam, the Philippines and Malaysia in their own EEZs, to the Chinese displacing Filipino fishermen from their traditional waters in the Scarborough Shoal, and to Chinese diplomats stonewalling on creating an official code of conduct in the South China Sea, China has been the antithesis to America's idea of the FON. Just looking at recent presidents, Obama did four FONOPS and Trump did 26,[25] Biden did 22. China has been forcing America's hand when it comes to keeping the South China Sea free for navigation.

---

25 Pedrozo, Raul (Pete) (USN [Ret.], is the Howard S. Levie Professor of the Law of Armed Conflict at the Stockton Center for International Law, US Naval War College. After his retirement from the U.S. Navy he was a Professor of International Law at the Naval War College and Principal Deputy Staff Judge Advocate, U.S. Indo-Pacific Command). "Does the Revised U.S. South China Sea Policy Go Far Enough?.":"Does the Revised U.S. South China Sea Policy Go Far Enough?" by Raul (Pete) Pedrozo (usnwc.edu). *International Law Studies* Vol.99 2022. Published by the Stockton Center for International Law in North Carolina, U.S.A. 2022. 74, 77.

What makes the rise of China dangerous is the belligerent nature of the PRC government when it comes to their claims in the South China Sea. Carolyn Bartholomew and Robin Cleveland bring up how the Chinese government reinforces its rule both at home and abroad. The authors write that Chinese leaders tell their own citizens and the world at large how China rose out of poverty, yet the official narrative omits international concerns that result from the repression of Xinjiang and Hong Kong, increasing debt, demographic decline and environmental degradation.[26] China doubles down on not talking about these issues, as the Chinese government uses ultranationalism to not only cover up for their failings back home but also to get closer to their goals. Examples of this behavior include how China bullies other countries to stop forming relations with Taiwan, and how China seeks to force an assimilation (or "Sinicization") onto minority groups in their country. China used coercion to demand Guyana cancel plans for a new Taiwan trade office, and ceased all rail shipments to Lithuania after that nation opened a similar office. In fact, when Lithuania became the first (and so far, only) European nation to formally use the name of Taiwan rather than Taipei, this led to the Lithuanian ambassador to China being expelled, and their own ambassador to Lithuania being recalled. Worse yet, Xi Jinping warned that anyone who attempted to "bully" China would "crack their heads and spill blood on a Great Wall of steel fortified by the flesh and blood of 1.4 billion Chinese people."[27] China's belligerence can no longer be ignored, for the Chinese government themselves do not want people to ignore that.

---

26 Bartholomew, Carolyn (Chairman of the U.S.-China Economic And Security Review Commission), Cleveland, Robin (Vice Chairman of the U.S.-China Economic And Security Review Commission). "2021 Report to Congress." (The Commission was created on October 30, 2000, by the Floyd D. Spence National Defense). :2021 Annual Report to Congress | U.S.- CHINA | ECONOMIC and SECURITY REVIEW COMMISSION (uscc.gov), U.S.-China Economic And Security Review Commission One Hundred Seventeenth Congress First Session November 2021. Published by U.S.-China Economic and Security Review Commission. Washington, D.C., U.S.A. November 2021. 1.

27 Ibid., 288, 290, 295, 306.

While the tension between China and Taiwan has not sparked into a full-on war, a war between the mainland and the island came close to reality nearly 30 years ago. Joel Wuthnow and Ash Carter wrote separately about what is known as the Taiwan Strait Crisis. Though Wuthnow described the Taiwan Strait Crisis happening from 1995-1996,[28] Carter pinpoints the fever pitch of the crisis in March 1996, when China fired missiles within 20 miles of Taiwan's coast. Carter said this move was a retaliatory action as Taiwan was holding its first democratic presidential election that same month.[29] Wuthnow explained that the crisis ended with America sending two carriers to the Taiwan Strait.[30] Carter summarized the move and the absence of a counteraction by the Chinese as then President Bill Clinton calling China's bluff. Afterward, Chinese Defense Minister Chi Haotian came to the Pentagon and said he "lost face" in this encounter. Then Secretary of Defense William Perry replied, "you deserved to."[31] The moves made by China and the United States show that the future of Taiwan is intrinsic to the future of the major powers of the South China Sea.

One factor that made the South China Sea such a prevalent topic in the past 20 years is that Southeast Asian countries rely on the sea for the economy and economic prospects in the future. Nandini Jawli, a freelance journalist, researcher, and a writer in Jaipur, India, spotlights

---

28 Wuthnow, Joel (Senior Research Fellow in the Center for the Study of Chinese Military Affairs, Institute for National Strategic Studies, at the National Defense University). System Overload: Can China's Military Be Distracted in a War over Taiwan? System Overload: Can China's Military Be Distracted in a War over Taiwan? (ndu.edu). Published by Center for the Study of Chinese Military Affairs Institute for National Strategic Studies National Defense University. Washington, D.C., U.S.A. June 2020. 7.

29 Carter, Ash (a late American government official and academic who served as the 25th United States Secretary of Defense from February 2015 to January 2017.) "Reflections on American Grand Strategy in Asia.":ReflectionsonAmericanGrandStrategyinAsia_0.pdf (belfercenter.org), Belfer Center Report. Published by Harvard Kennedy School Belfer Center for Science and International Affairs. Cambridge, Massachusetts, U.S.A. October 2018. 17.

30 Wuthnow. 7.

31 Carter. 17.

how the South China Sea provides the sustenance needed for countries surrounding the sea to survive. Jawli details how the majority of energy supplies for Japan, Taiwan, and China pass through the South China Sea. With China consuming 10 percent of the world's oil production and 20 percent of the world's energy, Jawli dubs the South China Sea as the "Second Persian Sea." The economic value of shipping, along with the sea's estimated oil reserves and fish stocks, has been estimated in the trillions of dollars. In that light, Chinese efforts to claim the entire sea because they discovered the islands in the sea 2,000 years ago, put China at odds with Vietnam, the Philippines, Malaysia, Indonesia, Brunei, and Taiwan. Not only that, but India is also wary of China controlling the South China Sea since 55 percent of India's trade passes through the South China Sea.[32] China has garnered the ire of their neighbors, for they seek not cooperation in the South China Sea, but domination.

Despite the South China Sea War being a war that would mostly be fought on sea, China's border with India would become another front in a war. This is shown by Sangit Ragi's description of the most recent (as of this writing) border clash between India and China. Ragi wrote about the June 15, 2020, battle of the Galwan Valley, where Chinese and Indian soldiers fought in hand-to-hand combat.[33] The combat was hand-to-hand, because both China and India had denied guns to their border troops as a way of ensuring that fewer people would die in possible border clashes like what happened in the Valley. Even with that restriction, Ragi mentions that 20 Indian soldiers and an

---

32 Jawli, Nandini (a freelance journalist, researcher, and a writer in Jaipur, India). "South China Sea and India's Geopolitical Interests." :South China Sea and India's Geopolitical Interests on JSTOR. *Indian Journal of Asian Affairs*, Vol. 29, No. 1/2 (June-December 2016). Published by the Indian Journal of Asian Affairs. Jaipur, India. 2016. 85-87, 89.

33 Ragi, Sangit (Sangit Kumar Ragi is currently a professor and Head in the Department of Political Science, University of Delhi.). *Indian Studies Review* Journal Of Centre For Study Of Politics And Governance Delhi Vol. 1 Number 2. July-December 2020 :Publication (cspgindia.com), Published by the University of Delhi. Delhi, India. December 2020. 2.

unspecified number of Chinese soldiers died in the battle. Ragi notes that the battle of Galwan Valley is reminiscent of the clash of Nathu La in 1967, where 80 Indian and 300 Chinese soldiers died from gunfire and artillery. This border conflict helped bring India closer to China's biggest rivals, such as America, Japan, and Australia forming the QUAD or the Quadrilateral Security Dialogue.[34] This border conflict can shape into a land front in addition to the larger sea war.

Another danger in any South China Sea war is that economic considerations would force countries to contribute. B. R. Deepak, in the 2018 book *China's Global Rebalancing and the New Silk Road*, brings into focus how China has wrapped countries around its finger economically. Deepak writes that between the 26th and 29th of March 2015, China revealed the Belt and Road Initiative (BRI) at the Boao Forum in Hainan, China.[35] The BRI is an economic and infrastructure project started by China to further connect itself with Asia, Africa, and Europe. As Deepak explained, 70 countries from said continents approved of this initiative.

China has identified six economic corridors along the BRI: the Bangladesh, China, India, Myanmar Economic Corridor (BCIM); the China Pakistan Economic Corridor (CPEC); the New Eurasia Land Bridge; the China, Mongolia, Russia Economic Corridor; the China Central Asia economic corridor; and the China Indochina Peninsula Economic Corridor.

As of 2018, Deepak shows that trade volume and investment in the BRI had exceeded $3 trillion and $50 billion, respectively. It was estimated that by 2020, China was expected to invest $1.4 trillion into the BRI, and that the economic value of the BRI could reach to $21 trillion. Deepak explains that India did not support the BRI since it

34 Ragi. 2, 37.

35 Deepak, B. R. (Professor B. R. Deepak was trained in Chinese history and India-China relations at the Peking University and Chinese Academy of Social Sciences, Beijing, at Jawaharlal Nehru University, New Delhi and University of Edinburgh, UK.), *China's Global Rebalancing and the New Silk Road* (springer.com), Published by Springer. New York, New York, U.S.A., January 22, 2018. V.

passes through Pakistani Kashmir, a territory that is disputed between India and Pakistan.[36] China's economic hold on the world creates the sense that whoever would win the South China Sea War, most of the world would lose.

China's aggression in the South China Sea has helped connect two of their biggest rivals: the United States of America and India. Jeff M. Smith brings up how the Indo-American relationship has been shaped in recent years. Smith explains that India sees its relationship with America as a necessity. In 2018, the Indian chief of naval staff, Admiral Sunil Lanba, bluntly admitted that "We can match what forces China can bring to bear in the [Indian Ocean Region]. But in the South China Sea, the dice is loaded in their … favor." Smith moves into 2020 with a report by *The Wall Street Journal* saying, "New Delhi is comfortable with the US maintaining its strategic base at Diego Garcia, a British territory island located about 1,100 miles southwest of the southern tip of India. It is steadily intensifying its military and diplomatic cooperation with the US, France, Australia and Japan—all nations that share New Delhi's concerns about China's attempts to establish itself as Asia's dominant power."[37] China's adamant behavior over the South China Sea proves to harm China's relations with major nations like the United States and India, as those nations unite to curb Chinese aggression.

Japan's relationship with America is another that has been strengthened as a result of China's ambitions. Yoshikazu Kato explains that what America, Japan, and China share in common is that they have the biggest economies in the world. He goes on further, writing, "The United States, at the moment, is still the most established superpower in almost all areas, including in the economy, military,

---

36 Deepak. V, 6-7.

37 Smith, Jeff M. (Research Fellow for South Asia in the Asian Studies Center in Pittsburgh, Pennsylvania). "UNCLOS: China, India, and the United States Navigate an Unsettled Regime." :BG3608.pdf (heritage.org). Backgrounder No. 3608 | April 30, 2021, Asian Studies Center. Published by the Heritage Foundation, Washington, D.C., U.S.A. April 30, 2021. 16, 18.

technology, education, soft power and so on. The US has also absorbed immigrants and large amounts of capital from all over the world by relying on the attractiveness of its institutions and value-systems, such as liberal democracy, rule of law, civil society, and individualism … China, on the other hand, is the greatest emerging power, as well as the largest developing country; in the country's official word, it has been trying to surpass the United States in a number of areas, including in the economy, military, and even soft power and regional and global influences." Kato elaborates while America would like to deepen its strategic partnership with Japan, China wants to co-opt the Japanese economy to challenge America's presence in the Asia Pacific.[38] Despite China's meteoric rise, America and Japan remain steadfast in their alliance.

With China's ambitions meeting friction in recent years, this has only fueled the Chinese government to look at their regional neighbors as enemies. From bullying naval vessels, to refusing to negotiate with neighboring countries, China has been a source of the majority of the tension in the South China Sea, as the Chinese government has claimed the region for itself. The recent history of the South China Sea can be best described as a boiling pot of overlapping land claims, naval exercises, and a growing need of natural resources.

---

38 Kato, Yoshikazu (an adjunct associate professor at the Asia Global Institute, University of Hong Kong. He is also a research fellow at the Charhar Institute, and columnist for *The New York Times* [Chinese edition]), "US-Japan-China Trilateral Relations: How their dynamics will shape Asia Pacific." :asiaglobal-papers-us-china-japan-relations.pdf (hku.hk). *Asiaglobal Papers: Working Paper*. Published by Asia Global Institute, The University of Hong Kong. June 2019. 4-5.

Chapter 2: China's Efforts at Rejuvenation

China's ability to rebound from its economic slump in the 1970s and become one of the biggest economies in the world is nothing short of a miracle. The Chinese government has benefited from their international reputation for the mass production of goods at a reliable pace. While China has become an economic hub for the world, the aggression of the Chinese government in the South China Sea has soured relations with most of its neighbors. This has garnered concern of the United States, Japan, India, and Australia, who are China's biggest rivals.

To understand China's focus on rejuvenation, it is important to understand Chinese history and the Chinese government's view on that history. Christian Wirth explained this in a 2020 journal article. Wirth noted that when Xi Jinping visited the Road of Rejuvenation permanent exhibition, he reportedly paid special attention to exhibits on the First Opium War (1840–2) and to charts illustrating the humiliating history of "how the West had occupied China's territories, established concessions and carved-out spheres of influence." Xi later went on to say, "the Chinese nation had suffered unusual hardship and sacrifice in the world's history … (and that) the Chinese people have never given in, have struggled ceaselessly, and have finally taken hold of their own destiny and started the great process of building the nation." This mentality of rewriting past wrongs fuels China's expansionist nature, including acts such as militarizing features in the South China Sea, which have angered Vietnam, the Philippines, America, Japan, and Australia. However, as Wirth mentions, China justifies its moves in the South China Sea as necessary to overcome "national humiliation." This is not helped by Chinese leaders claiming that China has a "peace-loving" character and that the country "never engaged in colonialism or aggression" in its 5,000-year history.[39] China's ability to revise history and turn said history into propaganda is how the Chinese government seeks to drum up support at home for the expansionist efforts.

The Chinese government focuses on asserting dominance on minority-led regions in their country, much as they assert themselves on the South China Sea islands. Richard Q. Turcsányi writes that domestic instability in Tibet and Xinjiang is linked to the Chinese government's hostile reactions to foreign nationals within their own nation. China's assertiveness has stemmed from the country's technological capabilities and to the fact that the 2008 financial crisis did not affect China as much as it affected the rest of the world. Another reason is China's increasing level of nationalism. As mentioned before, the Chinese government has used ultra-nationalism to popularize the idea to their people that the South China Sea is theirs. They use that same method to justify their actions against minority groups in China.

Turcsányi relates that China submitted the nine-dash-line plan to the UN in 2009, and that, since the 1990s, the Chinese have increased their naval activity in the South China Sea. He then brings up how China acts with its neighbors in the South China Sea and how it tries to take over the region. In 2011 and also in 2012, China repeatedly cut cables on Vietnamese seismic survey vessels in an apparent attempt to continue disrupting its oil-related activities. China rejected claims of any wrongdoing on its part, and asserted that Vietnam escalated the dispute by operating in the area belonging to China.

Between April and June 2012, the Philippines and China were

---

39 "Emotions, international hierarchy, and the problem of solipsism in Sino-US South China Sea politics." *International Relations*, Volume 34 Issue 1, March 2020. : Emotions, international hierarchy, and the problem of solipsism in Sino-US South China Sea politics - Christian Wirth, 2020 (sagepub.com), Published by Inquires Journal. Boston, Massachusetts, U.S.A. September 15, 2019. 5, 25-26, 31, 34, 38.

engaged in a standoff in the disputed Scarborough Shoal. The incident began on April 10, when Chinese fishing boats were spotted in the waters by Filipino reconnaissance planes. The crew of the Philippines' *Gregorio del Pilar*, the biggest warship of the navy, allegedly inspected the boats and discovered protected maritime species on them. When they tried to arrest the Chinese fishermen, they were prevented from doing so by accompanying Chinese surveillance vessels. The Chinese version of the story is different, though. China claims that the Chinese fishermen tried to take shelter in the shoal during harsh weather, and that the Philippine naval vessel blocked the entrance to the lagoon and harassed them. The Chinese surveillance vessels were then sent to protect the fishermen, according to the Chinese version of events. Even more importantly, China used economic measures to achieve its political goals.[40] China decided to cut its banana imports from the Philippines, officially citing quality issues. This resulted in a significant decrease of Philippine bananas in the Chinese market, both in terms of quantity and in terms of their share in the market. In their place, the imports of bananas from Ecuador to China surged. It is important to note that China and the Philippines have been butting heads over Scarborough Shoal since 1999.

On May 2, 2014, China deployed the Haiyang Shiyou 981 oil rig in disputed waters near the Paracel Islands, on the Vietnamese side of the median line between the Chinese and Vietnamese coasts in what China claims as its EEZ generated by the Paracel Islands. Thirty Vietnamese boats tried to intervene, but they were prevented from doing so by more than 80 vessels of the China Coast Guard. A year later, in 2015, China redeployed the oil rig near the Paracel Islands, although this time it was within China's undisputed waters and closer to China's coast. One of the direct outcomes of the incident was massive anti-China protests

---

40 Turcsányi, Richard Q. "Chinese Assertiveness in the South China Sea: Power Sources, Domestic Politics, and Reactive Foreign Policy": 978-3-319-67648-7.pdf (springer.com). Published by Springer. New York, New York, U.S.A. 2018. 3-4, 10, 12. 43-50.

in Vietnam, which led to damages to Chinese property and deaths of Chinese nationals. Similarly, this event contributed to Vietnamese strategic considerations and further deepening of Vietnamese relations with the USA and other actors, possibly balancing China's power.

Since 2014, China engaged in massive reclamation projects and the construction of artificial islands in at least seven locations in the Spratly Islands and at least three locations in the Paracel Islands. In the Spratly and Paracel Islands, the projects consisted of enlarging the dry area by reclamation works, building seawalls, and construction of various structures, including airfields, multi-floor buildings (some of them as tall as 10 floors), radar towers, gun emplacements, ship docks, and helicopter bases. Some posts were constructed in such a way as to be able to serve as air and naval bases of China and to be able to harbor the biggest Chinese naval vessels and all types of China's aircraft.

China's militaristic moves on the South China Sea can be seen as threats to other nations in the region. Going back to Fanell's "China's Global Naval Strategy and Expanding Force Structure: Pathway to Hegemony," China has invested a significant amount in ballistic missiles, satellites, and fiber-linked command centers to expand their means of dominance. Ever since 2000, the Chinese have planned to build up their blue water navy. The PLAN and the Chinese Coast Guard (CCG) have increased their military operations in China's first and second island chains since 2000.

During the 19th National Party Congress in 2017, President Xi Jinping emphasized the idea of Chinese rejuvenation. This idea of rejuvenation is linked to the expansion of the capabilities of the navy. China's goal for the PLAN is to assert their economic and security interests in the coming decades, to enjoy a building cost advantage over their rivals, and to have technological advances to overcome future problems. Fanell mentions that President Xi has restructured the PLA in the seven military regions into five theater commands. Xi has also reorganized the Central Military Commission (CMC) by establishing and subordinating the army's service headquarters, strengthening CMC's role in strategic missile, air, and naval forces.

Xi also established the Strategic Support Force (SSF) to integrate space, cyber, and electronic warfare capabilities into the military. Fanell explains that the Chinese government plans to use the PLAN to help secure China's global interests, especially in their $1.6 trillion Belt and Road and Initiative (BRI). The PLAN expansion from 2000-2018 exceeds any national buildup post-WWII except for the US Navy in the '80s under President Ronald Reagan. Fanell brings up a report by the US Office of Naval Intelligence (ONI) that suggests the PLAN's military exercises during the past decade were designed to map out how to perform multiple missions for naval combat against naval forces. It's hypothesized that PLAN assault forces will be led by surface strike groups composed of the service's premier combatant, the Type 052D Luyang III-class guided-missile destroyers. The Type 054C Luyang II-class guided-missile destroyers, Type 054A Jiangkai III-class guided-missile frigates and Soviet-built Sovremenny-class destroyers would also likely take part. With the superior arsenal of antiship cruise missiles (ASCMs) these strike groups can provide support for both naval and ground battles.

Since 2012, the PLAN has been exercising around the Senkaku Islands, which are owned by Japan. Fanell reveals that the People's Liberation Army Air Force (PLAAF) has engaged in exercises around Taiwan with planes capable of holding nuclear bombs. China's naval firepower will come from submarines armed with supersonic, sea-skimming, 290 nm-range YJ-18 ASCMs, as well as ASCMs from the PLAAF. Fanell warns that if China were to strike first, America and her allies would be left unprepared to defend the Senkakus.

Despite Taiwan being a more difficult target, China now has the capabilities to launch a no-warning attack on the island nation. China now has treaty ports in Southeast Asia, Africa and Europe to control the Malacca Strait and the Suez Canal. In early 2017, the commander of the US Pacific Forces warned Congress that the PLAN was expanding in presence and influence thanks to the commercial network created from the BRI. The PLAN is able to do this because the Chinese government is using state-owned companies and politically linked private firms to create facilities with the capacity to provide logistic support to PLAN warships. This has been called the "first civilian, later military" approach, using China's access to treaty ports and their economic capabilities to enhance those ports' ability to support the PLAN. China continues to build and train their naval and amphibious forces to prepare for future military campaigns, using amphibious dock landing ships, air-cushion landing craft, and shipborne helicopters for their exercises.

This led to the expansion of the People's Liberation Army Marine Corps (PLAMC) to 100,000 personnel. This new PLAMC will be placed in several locations outside of China such as Gwadar, Pakistan, and the new PLAN base in Djibouti. This move is intended as a thorn in the side of the Americans and their allies, and a boost to China's allies such as Pakistan, Iran, and Russia. As of 2015, the PLAN has 56 amphibious warships, ranging from a few WW2-era ships to the new Yuzhao-class type 071 amphibious transport dock ships which can carry helicopters and armored vehicles along with troops. China's goals for its 2030 fleet are to be able to conduct near-seas active-defense operations, far-seas operations, "goodwill" deployments, surge operations, carrier strike-group operations, amphibious assault group operations, and submarine-launched ballistic-missile patrols. In December 2012, a Chinese maritime surveillance plane entered the airspace of the Senkaku Islands in the first time in 50 years.

In 2017, the PLAAF increased its presence near Japanese airspace. The PLAAF routinely sends bombers to threaten Japan, Guam, and the nations within ASEAN. This failure of America and Japan to defend in the Senkaku Islands may give China a sign that the Senkakus are weakly defended and could be an attractive target. China could surprise with an air attack. China's H-20 bomber would be on par with America's B-2 Stealth bomber. China has put more emphasis on rocketry with the creation of the People's Liberation Army Rocket Forces (PLARF). Coordinated and synchronized strikes could employ antiship ballistic and cruise missiles, land-attack cruise missiles and air strikes with precision-guided munitions. Counter-C4ISR (command, control, communications, computers, intelligence, surveillance, and reconnaissance) systems would also be used. The goal would be to take either Taiwan or the Senkakus and isolate the American, Japanese, and Taiwanese militaries.

The information sector is also getting reorganized to work alongside the military. A planned five-year effort to reform the PLA Informization department began in 2015. The mission of the Strategic Support Force (SSF) is focused on support for psychological, electronic, cyber, and space warfare. The two organizations are responsible for informational warfare, the Third Department of the PLA General Staff Headquarters (3PLA) and the Fourth Department (4PLA). By combining cyber-attacks, misinformation, disinformation, and psychological operations (Psyops) with conventional warfare, China plans to distract the enemy while engaging the targets. Deng Xiaoping advised the Chinese government to hide their military assets and bide their time. To Xi Jinping, that time has come. China wants to confuse, demonize and demoralize America.

On February 17, 2017, Xi Jinping called for unity with the 60 million Chinese people living overseas. The Xi regime wants to use these people to undermine military and political opposition to the CCP and advance the military and political power of the CCP to the Chinese Dream. In 2014, the PLA established a permanent joint operations command (JOC) center to integrate the operations of the Army, Navy, and Air Force. China has even gone as far as the Baltic Sea. China has treaty ports with Australia, Cambodia, Malaysia, Indonesia, Brunei, Myanmar, Bangladesh, Sri Lanka, Pakistan, Djibouti, Tanzania, Mauritius, Namibia and Greece, and is negotiating for ports with the Maldives, Scandinavian countries, and Greenland.

Since 2013, the PLAN has deployed nuclear submarines to the

Indian Ocean to threaten India. In 2016 a PLAN warship took an American underwater glider, a brazen theft of technology by the Chinese military. Since 2012, the PLAN has conducted joint exercises with the Russian navy called JOINT SEA. The first exercise took place in the Yellow Sea in 2012. In 2017, China sent three warships to the Baltic Sea for several weeks of joint exercises with 10 Russian warships. This follows years of pointedly declining to denounce the Russian seizure of Crimea and of ignoring Estonia's complaints that the Russian army and navy are operating too close to Estonian borders.

The United States has failed to make a proper response to the advancement of the PLAN. America has not responded to the several violent acquisitions of South China Sea Islands over a period of nearly 50 years. This lax attitude on Chinese encroachment has led to nations like the Philippines letting China militarize the islands around them. The war will not just be fought by the military, but also in the hearts and minds of the people in the Indo-Pacific. Australia, America's ally, has sold a port to China in Darwin for 506 million Australian dollars (AUD). China is looking at French Polynesia as a future stepping-stone for future endeavors in Antarctica. China has invested a third of a billion US dollars to build an agricultural project in French Polynesia. From 2005-2014, China gave almost $7 billion worth of loans to Sri Lanka. In 2013, the Maldives elected a pro-China president, giving China a foothold on two strategically important countries in the Indian Ocean. In 2015, China state media stated that the ports of Kyaukpyu in Myanmar, Chittagong in Bangladesh, Colombo in Sri Lanka, Aden in Yemen, and the ports of the Maldives are potential industrial hubs to support the PLA military operations. China uses debt traps in Sri Lanka and Myanmar to gain leverage in these countries.

In 2015, the Chinese government announced plans to invest $60 billion into Africa. In July 2017, China established a military logistics base in Djibouti. China buttered Djibouti up with a $600 million port terminal, a $4 billion terminal for liquefied natural gas imports, a $600 million deal for two new airports, and a $4 billion railroad. In the same month, China conducted a military exercise on their base in Djibouti. This was not an anti-piracy exercise, but one intended to demonstrate China's establishment in the Horn of Africa. In May 2018, China used laser technology, either from the Djibouti base or a nearby PLAN ship, against two United States Air Force pilots. Both suffered minor eye injuries. In February 2018, Djibouti terminated use of the treaty ports by America and its allies. The supplies at these ports could be given to China. China also used the World Bank to fund their project in the Tanzanian port of Dar es Salaam in 2017.[41] China's military buildup is not only increasing the size of their army but also resulting in the encirclement of Indo-Pacific with military installations and ports.

In addition to its military, diplomatic, and economic strategies, the People's Republic of China also seeks to further its ends through the use of Digital-Age information technology, as Dennis F. Poindexter elaborated on in his 2018 book. Poindexter claims that with information warfare, a government can capture the minds of the population, influence its enemies and engage with allies in a way not previously understood. Even talking about information warfare is controversial. Poindexter brings up an example with the conviction of US General James Cartwright for disclosing to *The New York Times* information on the development of Stuxnet, a computer worm that could destroy the centrifuges used to make nuclear material in Iran. Though General Cartwright was pardoned, this shows how important information warfare is to the US government.

Where China comes in is that Chinese officials have talked about using that form of warfare in a scenario of war. With America having information weapons like Stuxnet in 2016, a war between America and China could be the first major conflict in which information warfare is used. What makes China's entrance into information warfare

41 Fanell, James E. "China's Global Naval Strategy And Expanding Force Structure: Pathway to Hegemony." :CHINA'S GLOBAL NAVAL STRATEGY AND EXPANDING FORCE STRUCTURE on JSTOR, *Naval War College Review*, Vol. 72, No. 1 (Winter 2019), pp. 10-55, Published by the U.S. Naval War College, Newport, Rhode Island, U.S.A. 2019. 14-22, 24-31, 33, 35-38.

interesting is that it fills the niche that the Chinese government was looking for. This is brought up by Taylor M. Fravel, who quoted then Chinese President Jiang Zemin in 1993 saying, "we must clearly place the basis of preparations for military struggle on winning local wars under informationized conditions." The "informationized conditions," Fravel explains, are an indication that this type of warfare is becoming more integrated within the Chinese military and Chinese strategy. This includes nonlethal tactics such as information, cyber, electronic, public opinion, psychological, and legal warfare. This change came about due to China's observations of American military actions in the 1990s and the 2000s. When America came to the aid of Kosovo during the Yugoslav Wars (1991-2001), the Chinese saw this as a sign that America would aid Taiwan in the event of an attack. Another example Fravel brings up is that during the first 9 months of the Iraq War (2003-2011), China looked closely on the combination of information warfare with more conventional weaponry.[42] This shows that China has had decades worth of time developing their own tactics on information warfare.

One of Xi Jinping's greatest reformations was the merging of civilian research and the military. This is showcased in a journal article by Amrita Jash, who notes that following Xi's rise to power, the Military-Civil Fusion (MCF) became one of the major additions to China's national strategy. The MCF is the integration of civilian research and commercial sectors with the military and industry. This project aims to be the tool which China will use to catch up to the speed and scope of the West's technological advancements. China wishes to use the MCF to foster a fusion of the military and national defense with the country's economic and social systems. In November 2013, Xi Jinping said at the Third Plenum of the 18th Central Committee in Beijing that "We must deepen military structural and personnel

42 Fravel, M. Taylor. *Active Defense: China's Military Strategy Since 1949.*:Active Defense: China's Military Strategy Since 1949 - M. Taylor Fravel - Google Books, Published by Princeton University Press. Princeton, New Jersey, U.S.A. 2019. 217-219, 223, 226.

allocation adjustment and reform, move adjustment and reform of military policies and institutions forward, and promote the integrated and deep development of the military and the people."

Jash explains that two years later, Xi said his goal was to make a strong military with Chinese characteristics. Specifically, he wanted to optimize the military structure, enhance the military-civilian integration, and build a modern military with Chinese characteristics that can win a digital war. This resulted in the most ambitious restructuring since the conception of the PLA in 1949. In June 2017, Xi expanded on his MCF strategy: "We must accelerate the formation of a full-element, multi-domain, and high-return military-civil fusion deep development pattern, and gradually build up China's unified military-civil system of strategies and strategic capability." Xi's motivation for the MCF strategy is the need to modernize China's defense science and technology to meet the demands of winning local wars in the South China Sea. Jash claims that China learned by observing the Gulf Wars and the Kosovo crisis that technology is a key to modern warfare, and that understanding is pushing China to upgrade its defense technology. As examples of this new priority, Jash cites Yunzhou Tech, a leader in unmanned surface vessels, Ziyan, a major company in drones and unmanned helicopters, and Kuang Chi Technologies, which applies machine learning to research on military meta-materials.

According to Jash, the MCF strategy encompasses five efforts:

1) Fusing China's defense industry and civilian technology with the commercial industry.
2) Integrating science and technology innovations across the military and civilian sectors.
3) Cultivating talent and combining military and civilian knowledge.
4) Building military needs in civilian infrastructure and using civilian construction for military purposes.

5) Expanding China's national defense mobilization system to include all relevant aspects of its society and economy for use in competition and war.

Jash concludes that China wants to have a world class military by 2049.[43] This integration of military and civilian technology is intended to help China with future conquests in the South China Sea and to become the world's superpower.

While the modernization of the Chinese military has been spotlighted by Western media in recent years, this project actually was in the works for several decades, as Andrea Ghiselli explains in his 2021 book. Ghiselli writes that China's military modernization has been going on for 35 years. He notes that the 1985 Central Military Commission, led by then Chinese President Deng Xiaoping, declared that China was not going to face a major war or nuclear war as in the times of Mao, but rather local and limited wars.[44] This provides context that the Chinese have planned to attain the status of superpower for more than three decades.

Taiwan has not only become the biggest point of contention between America and China, but also a hindrance to China's short- and long-term goals. Dario Kuntic explains that China sees an independent Taiwan as a source of internal and external problems. Not only is the Taiwan situation a last vestige of humiliation for China, but it may provide inspiration to independence-minded groups in Tibet and Xinjiang. This is especially true when America praises Taiwan

---

43 Jash, Amrita (Research Fellow at the Centre for Land Warfare Studies, New Delhi, India). "China's military and civil fusion.": Chinas-Military-Civil-Fusion-Strategy-Building-a-Strong-Nation-with-a-Strong-Military.pdf (researchgate.net). *CLAWS Journal* Winter 2020, Published by the Centre for Land Warfare Studies (CLAWS). New Delhi, India. December 2020. 42-47, 54.

44 Ghiselli, Andrea (Assistant Professor at the School of International Relations and Public Affairs at Fudan University [China]). *Protecting China's Interests Overseas: Securitization and Foreign Policy*. :Protecting China's Interests Overseas: Securitization and Foreign Policy - Andrea Ghiselli - Google Books. Published at Oxford University Press. Oxford, Britain. 2021. 49.

for being the only Chinese country where the people enjoy a republic that upholds the rights of its people. Outside of China, Taiwan along with Japan, Malaysia, Indonesia, the Philippines, and Australia make up what is known as the first island chain, which could impede the PLAN's movement in time of war. If China were to take Taiwan, the PLAN would be free to strike deeper into American territory.[45] To the Chinese government, Taiwan poses as the biggest thorn in their side, in terms of their plans.

Along with focusing on external issues, China also looks at fixing internal issues as part of the rejuvenation effort. Michael Clarke shows in his journal article that minority groups within China are affected by China's ultra-nationalistic policies. Though Xinjiang has only recently been of great interest for the international community, Clarke explained the Chinese government has long been sending ethnic Han Chinese people to Xinjiang to dominate the region over the native Uyghurs, a majority Muslim population in Southwestern China. However, the Chinese government has doubled their efforts to control the region as China has also created laws to crack down on Uyghurs practicing their religion. Uyghurs risk being sent to political education centers to correct their "Deviant" behavior. China frames key elements of Uyghur culture as pathologies to be "cured." The Chinese government describes these efforts as counterterrorist measures and justifies the use of re-education camps as a form of preventive counterterrorism.[46]

---

45 Kuntic, Dario (Minister Counselor at Permanent Representation of the Republic of Croatia to the European Union). "The Ominous Triangle: China-Taiwan-the United States relationship.": The Ominous Triangle: China-Taiwan-the United States relationship (srce.hr). *Croatian International Relations Review* 21. Published in Ministry of Science and Education of the Republic of Croatia. Zagreb, Croatia. March 2015. 247-248, 252-253.

46 Clarke, Michael (Associate Professor at the National Security College, Crawford School of Public Policy, Australian National University (ANU) and Director of the ANU-Indiana University Pan-Asia Institute.), January 2019, "China's Applications of the Three Warfares: China's Application of the 'Three Warfares' in the South China Sea and Xinjiang." (anu.edu.au), Orbis Published for the Foreign

China's treatment of the Uyghurs displays how far it will go to solidify control over the Southwestern region.

Though China has been aggressive in the South China Sea since the latter half of the 20th century, such aggression escalated in the mid-to-late 2000s. Andrew Chubb, in a 2021 journal article, describes post-2007 Chinese policy on the South China Sea as the administrative buildup in the region and the increased use of coercion by the Chinese government. Chubb goes on further with Chinese coercion by writing that four types of assertive actions in maritime and territorial disputes are identifiable, based on their increasingly serious implications for the positions of rival claimants:

1) Declarative: verbal assertions via non-coercive statements, diplomatic notes, domestic legislation and administrative measures, or international legal cases.
2) Demonstrative: the unilateral administration of disputed possession, including patrols, surveys, resource development, construction of infrastructure, state-sanctioned tourism or activism, domestic judicial proceedings, and cooperative agreements with third parties.
3) Coercive: a threat or imposition of punishment, which may be verbal, diplomatic or administrative, economic punishment, warning shots, or physical interference with foreign activities in disputed areas.
4) Use of force: the application of military force or direct seizure and occupation of disputed possessions.

In his examination, Chubb provides some examples. For instance, Japan launched legal proceedings against a PRC fishing boat captain who ordered the ramming of a Japanese Coast Guard ship in the disputed territorial seas around the Diaoyu/Senkaku Islands in 2010.

Policy Research Institute by Elsevier Ltd. Amsterdam, Netherlands. Spring 2019. 14, 16-17, 19.

The use of force can be seen through events such as the clash in 1988 between China and Vietnam over the Paracel Islands, and in the battle of Galwan Valley in 2020. Coercive behaviors are those involving the threat or use of punishment against an adversary. Coercive actions pose more serious risks to stability than declarative or demonstrative moves because they present a relatively narrow set of choices to other parties: alter their behavior or continue and risk incurring punishment.[47] China's attempt to take the South China Sea is a multi-pronged affair that has only flared up in the past 15 years.

While China is preparing for a war in the South China Sea, the government is working on motivating its populace to support that war. This is brought up by Jordan M. Sandy, who used the term "patriotic education" to describe the curriculum China uses in an effort to turn children into hyper-nationalists. That education is centered around the "Century of Humiliation" or the "Century of National Shame," which is the center point of Chinese nationalism. The "humiliations" refer to the major conflicts the Chinese fought from 1839-1945, including two Opium Wars, the Sino-Japanese Wars, and the Boxer Rebellion. The patriotic educational program started shortly after the Tiananmen Square Massacre. This curriculum was in place to counter any more democratic protest, by controlling what their people were taught. The purpose of the "Century of Humiliation" narrative is to showcase the failures of the Qing Dynasty and Nationalist China, and the victories of Communist China. More specifically, victories over foreign oppressors. China's version of nationalism is different from how Western countries view nationalism. China views nationalism as loyalty to the state that creates a sense of pride and hope in the Chinese

---

47 Chubb, Andrew (British Academy Postdoctoral Fellow in the Department of Politics, Philosophy, and Religion at Lancaster University.). "PRC Assertiveness in the South China Sea: Measuring Continuity and Change, 1970–2015.": PRC Assertiveness in the South China Sea: Measuring Continuity and Change, 1970–2015 | International Security | MIT Press. *International Security* (2021) 45 (3). Published by MIT Press. Cambridge, Massachusetts. 2021. 83, 87-90, 98.

people,[48] whereas Europeans view nationalism as a state loyal to its people. China's plan is to educate its youth, so they remain loyal to the state and willing to go to war in the South China Sea.

Chinese history, and the government's perspective on that history, also has consequences for people within China, especially the minorities in the country. Denny Roy, in his 2013 book, shows what China's focus on rejuvenation has meant for minorities. Roy writes that China has drawn three lessons from its history:

1) The Chinese deserved to be recognized as the world's greatest country, which has been true for most of their history.
2) Dividing the country impedes its greatness, thus lost land must be recovered and preserved.
3) China's history has proven that the other great powers of the world will only exploit the Chinese.

Roy adds that while Taiwan remains as the main concern for Chinese reunification, Tibet and Xinjiang are not far behind, because the Lamaist Buddhist and Islamic lifestyles respectively, clash with the Chinese government's atheistic beliefs.

Given the different cultures of the two regions, the peoples of Xinjiang and Tibet responded differently to China's crackdown on their cultures. While the Uyghur people of Xinjiang committed terrorist attacks from 2008 to 2010, Tibet's religious leader, the Dalai Lama, self-exiled to India in 1995. China responded strongly to both moves. In both situations, China began to crack down on the culture and religions of minorities in their western regions, with critics of China's counter-terrorist actions in Xinjiang calling it an excuse to persecute the Uyghurs. China arrested the Dalai Lama's successor (Panchen

---

48 Sandy, Jordan M. (M.A. School of Public and International Affairs at the University of Princeton in New Jersey). "Chinese Nationalism and the South China Sea.": Ohio LINK ETD: Sandy, Jordan M. Published by Wright State University. Fairborn, Ohio, U.S.A. 2020. 21-22, 31, 34.

Lama) and sent him back to Tibet. It does not help matters that the Tibetan government-in-exile accused the Chinese government of cultural genocide, citing the massacre of Tibetan monks and nuns as an attempt to make the Tibetans a minority in Tibet. Meanwhile, the Chinese government claimed the Tibetan Theocracy was backward, and that Tibetans were more prosperous under the CCP rule than they would be under the former Tibetan government. Chinese leadership has gone as far as label the Dalai Lama as a Separatist.[49] This does not only show how China's policies are hurting minorities but also presents a hint of how a Taiwan occupied by China would look.

While China's Belt and Road Initiative project has been praised for helping less-wealthy countries, skeptics see it as too good to be true. Tom Miller wrote in 2019 about the issues concerning the BRI. Miller said critics of the BRI look at it as a debt trap for smaller and less developed countries such as Pakistan, Sri Lanka, the Maldives, and Malaysia. Both Pakistan and Sri Lanka have economic partnerships and significant debts with China. In support of this claim Miller cites a 2018 report from the US think tank the Center for Global Development. Of 68 countries hosting BRI projects analyzed in the report, 23 were at risk of debt distress; eight countries (Pakistan and the Maldives included) were at a high debt risk; and all eight of these countries could default on their sovereign debt if these BRI projects come to fruition.

China has used the BRI to get leased ports in strategic locations including the Maldives and Malaysia, and to build ports in Myanmar, Sri Lanka, and Pakistan that can support their warships and submarines. These controversial projects led to the overthrow of the president of the Maldives, Abdulla Yameen, in September of 2018, and Prime Minister of Malaysia Najib Razak in May of 2018. The BRI projects in Malaysia

---

49 Roy, Denny (Ph.D. in political science, University of Chicago, 1991, writes mainly on Northeast Asian international security issues, particularly those involving China). *Return of the Dragon: Rising China and Regional Security.*: Return of the Dragon: Rising China and Regional Security - Denny Roy - Google Books. Published by Columbia University Press New York, New York, U.S.A. 2013. 15-18.

were scandal-ridden, with billions of dollars being siphoned out of a state fund. Malaysia's new leader Mahathir Mohamad suspended more than 20 billion US dollars' worth of unequal treaties made with Chinese firms. Mohamad said in August of 2018 at a Beijing press conference: "We do not want a situation where there is a new version of colonialism because poor countries are unable to compete with rich countries."

Since then, Chinese projects have been suspended in Mexico, Myanmar, and Indonesia. However, countries like Pakistan are more reliant on China than ever before. China's infrastructure diplomacy has been a strong arm for China's foreign policy. Despite diplomatic hiccups, Malaysia's biggest trade and investment partner is China. America's alliance with South Korea and Japan shows a major weakness in the Chinese strategy. Offers of development aid can lure developing countries into China's fold, but developed nations allied with each other and the United States are less likely to succumb to China's attempts at dominance over Southeast Asia. For all China's talk of a community, the country will have a hard time to get to America's sphere of influence.[50] While China's sphere is ever-growing, it has not had the time to develop the relationships that America has with countries in the South China Sea.

China has been at odds with America and its partners, but this conflict also affects countries in the South China Sea that want to remain neutral. This is brought up by Rabindra Sen, who noted in his 2018 journal article that, since 2013, China has been building artificial islands in the South China Sea. ASEAN's difficult position is obvious, given the comment by Xu Bu, the Chinese ambassador to ASEAN, who noted that China had been the biggest trade partner for ASEAN

---

50 Miller, Tom (managing editor of the China Economic Quarterly and writes about China and India at Gavekal Research, a global economic research service based in Hong Kong). *China's Asian Dream: Empire Building along the New Silk Road*: China's Asian Dream: Empire Building along the New Silk Road - Tom Miller - Google Books. Published by Zed Books Ltd. London, Britain. 2019. XIII-XIV, XVI, XVIII-XX, 15, 19.

for the past eight years. Said trade relationship was worth $452 billion in 2016. Despite this, also in 2016, former ASEAN Secretary General Le Luong Minh said that both China and America are important partners for ASEAN and the organization does not want to have to pick a side. While China will approve short-term agreements with ASEAN in terms of conduct within their waters, they will not enter into any agreement that would restrict them from claiming their rights in ASEAN waters.[51] This creates a difficult middle ground for ASEAN as they deal with one of the biggest nations in the region.

One other reason for tension in the South China Sea is the intrusion of Chinese fishing vessels in other nations' Exclusive Economic Zones (EEZs). This is brought up by Hongzhou Zhang, who writes that "According to The China Power project at the Center for Strategic and International Studies (CSIS) in Washington, D.C., of the forty-five major incidents identified in the South China Sea between 2010 and 2016, at least one China Coast Guard (CCG) (or other Chinese maritime law enforcement) vessel was involved in 71 per cent of incidents, mostly ones involving fishing vessels." Zhang explains that this is done to solidify China's claims of the South China Sea with a constant presence in these contested areas. Given that there are 90,000 Chinese fishing vessels, China has many opportunities for bullying tactics in contested regions in the South China Sea. However, this has triggered a response from South Korea, which, in 2012, seized 467 Chinese fishing vessels that were illegally operating in the South Korean EEZ. Zhang brings up that "Between 2011 and 2015, the (Chinese) central government's financial support to the marine fishery amounted to RMB146 billion

---

51 Sen, Rabindra (Professor at the Department of International Relations, Jadavpur University, Kolkata, India). "China's South China Sea Strategy: Balancing Triumphalism and Accommodation." IJEAS : *The International Journal of East Asian Studies* Vol. 7 No. 1 (2018).:China's South China Sea Strategy: Balancing Triumphalism and Accommodation | WILAYAH : The International Journal of East Asian Studies. Published by the University of Malay. Kuala Lumpur, Malaysia. December 15, 2018. 27, 31, 36,40.

(US $21.8 billion), nearly three times higher than during 2006-2010."[52] The use of fishing boats has given China a bigger presence in the South China Sea, but at the cost of souring relations with their neighbors.

Though the First Island Chain is a problem for the PLAN, the Chinese government is finding ways to circumnavigate this issue. The government is developing missile technology to better enforce its claims. This is explored by Murray Hiebert, Phuong Nguyen, and Gregory B. Poling in their journal article. The authors explained that since the mid-1990s, China has developed missile technology that is aimed at America's naval bases in the Western Pacific.[53] This would impede the Americans in their support for Taiwan and their troops in Japan and Korea.

Chinese investments have placed many of the region's smaller nations into debt, and China has exploited that debt situation to acquire leased ports in developing nations outside the first island chain, according to Brendon J. Cannon and Ash Rossiter. Cannon and Rossiter explain how China gains access to ports outside of China, by starting with the Maldives, an island nation in the Indian Ocean. They write that "Maldives' external debt will hit 51.2 percent of GDP in 2021, up from 34.7 percent in 2016. The Maldives also has a US$286 million trade deficit with China. Former president of the Maldives,

52 Zhang, Hongzhou, Bateman, Sam. "Fishing Militia, the Securitization of Fishery and the South China Sea Dispute." :Fishing Militia, the Securitization of Fishery and the South China Sea Dispute on JSTOR. *Contemporary Southeast Asia* Vol. 39, No. 2 (August 2017), pp. 288-314. Published by ISEAS - Yusof Ishak Institute, Singapore. August 2017. 289, 292-293, 299.

53 Hiebert, Murray (A senior fellow and deputy director of the Sumitro Chair for Southeast Asia Studies at CSIS), Nguyen, Phuong (A research associate with the Sumitro Chair for Southeast Asia Studies at CSIS), Poling, Gregory B. (director of the Asia Maritime Transparency Initiative and a fellow with the Sumitro Chair for the Southeast Asia Studies at CSIS at Washington, D.C.). "Examining The South China Sea Disputes." :Examining the South China Sea Disputes: Papers from the Fifth Annual CSIS South China Sea Conference (csis-website-prod.s3.amazonaws.com). Papers from the Fifth Annual CSIS South China Sea Conference: A Report of the CSIS Sumitro Chair for Southeast Asia Studies. Published by the Center For Strategic & International Studies, Washington, D.C., U.S.A. September 2015. 4.

Mohamed Nasheed claims that 80 percent of the debt the Maldives has is from China, and that China would make the Maldives cede territory. He added that China already owned 16 islands in the country and '(w) ithout firing a single shot, China has grabbed more land than the East India Company at the height of the 19th century.'"

The two authors go further by writing that "having acquired leasing rights to Pakistan's Gwadar port for 40 years, Greece's Piraeus port for 35 years, sections of Djibouti's port for ten years, Sri Lanka's Hambantota port for 99 years, 20 percent of Cambodia's total coastline for 99 years, and the Maldivian island of Feydhoo Finolhu for 50 years, Beijing is now pressuring Myanmar to raise China's stake from 50 percent to 75 or 85 percent in the Kyaukpyu port on the Bay of Bengal, and to lease it for 99 years as well—at least if Myanmar does not want to pay a penalty for reneging on the US$3 billion energy dam deal in Myitsone."[54] China's use of economic coercion has given it the modern-day equivalent of 19th-century treaty ports which could prove to be an issue if China utilizes the ports in the event of a war. China under Xi Jinping is restructuring the military and expanding its reach by using ports that belong to other nations. Jiacheng Li brings this to light in a 2017 journal article, in which he reveals that on Sept. 3, 2015, President Xi announced that the Chinese military would cut 300,000 personnel from the army to strengthen the air force, navy, and the strategic missile force, putting an emphasis on a variety of means to take on a multitude of threats. Li writes that China also has a presence in the Indian Ocean including the Chittagong port of Bangladesh, the Sittwe port and Coco Islands of Myanmar, the Hambantota port in

54 Cannon, Brendon J. (Assistant Professor of International Security at Khalifa University, Abu Dhabi, UAE), Rossiter, Ash (Dr. Ash Rossiter has been assistant professor of international security at Khalifa University, Abu Dhabi, UAE).: Rising-Powers-Quarterly-Volume-3-Issue-2.pdf (kureselcalismalar.com). *Rising Powers Quarterly*: The "Indo-Pacific" – Regional Dynamics in the 21st Century's New Geopolitical Center of Gravity Vol. 3, Issue 2|August 2018. Published by the Marmara University, Istanbul, Turkey. 2018. 71-73, 75.

Sri Lanka, the Gwadar port of Pakistan, and the Bagamoyo port of Tanzania. Since these are not bases, there are no grounds for America or India to protest these actions. In April of 2017, China was granted permission for 40 years to operate in the port of Gwadar in Pakistan, which western observers see as a "Chinese version of Gibraltar."[55] This gives China the capability to open a bigger theatre in the Indian Ocean and threaten the containment of Chinese expansion.

China's external ambitions may be hampered by internal obstacles. David Bandurski claims that although China's drive toward technological independence raises alarms in the West, there are internal conditions that could prevent said independence. One obstacle is the sense of stagnation within the Chinese population. The youth of China refuse the idea of a "continued struggle" that is promoted by the older party elites. He writes, "They opt instead for 'lying flat,' or tangping (躺平). The 'lying flat' movement calls on young workers and professionals, including the middle-class Chinese who are to be the engine of Xi Jinping's domestic boom, to opt out of the struggle for workplace success, and to reject the promise of consumer fulfilment. For some, lying flat promises release from the crush of life and work in a fast-paced society and technology sector where competition is unrelenting. For China's leadership, however, this movement of passive resistance to the national drive for development is a worrying trend—a threat to ambition at a time when Xi Jinping has made grand ambition the zeitgeist of his so-called 'New Era.'"[56] China's struggles with the lying flat movement demonstrate an underlying issue with Chinese society as the Chinese government is hyper-focused on becoming the

---

55 Li, Jiacheng (an associate professor at Liaoning University's School of International Studies in Liaoning, China). "China's Indian Ocean Strategy: Developing China's Indian Ocean Strategy: Rationale and Prospects." *China Quarterly of International Strategic Studies* Vol. 03, No. 04, pp. 481-497 (2017). Published by World Scientific Publishing Co Pte Ltd. Singapore. 489-491, 494.

56 Bandurski, David (the Co-Director of the China Media Project, a research program in partnership with the Journalism & Media Studies Center at the University of Hong Kong). "The 'lying flat' movement standing in the way of China's innovation drive." Brookings Institution. July 8, 2021.

most advanced nation in the world. This creates a generational split between the elders who want to continue the status quo, and the youth who see the problems with said status quo and are protesting any way they can in a country that prohibits dissent against the government.

Bandurski ties the beginning of the lying flat movement to a post on the Chinese social media site Baidu titled "Lying Flat Is Justice" in April 2021. In the post, the author explained their life after two years of unemployment. They explain that the stress of that "contemporary life" was unnecessary, and "the product of the old-fashioned mindset of the previous generation." They continue with "I can be like Diogenes, who sleeps in his own barrel taking in the sun." This post brought up further discussion on "lying flat" and gained traction, as others recounted similar experiences of being over-worked and over-stressed. Bandurski mentions that there has been pushback in China on the "lying flat" narrative. Wang Xingyu, an official at the China University of Labor Relations, writing in the *Guangming Daily*, criticized the movement by saying, "The creative contribution of our youth is indispensable to achieving the goal of high-quality development" and "Attending to those 'lying flat,' and giving them the will to struggle, is a prime necessity for our country as it faces the task of transitioning development."

Nanfang Daily, the mouthpiece of Guangdong's CCP leadership, expressed disgust at the movement, saying: "At any time, no matter what stage of development, struggle is always the brightest base color of youth," and "In the face of pressure, choosing to 'lie flat' is not only unjust, but shameful. There is no value whatsoever in this poisonous chicken soup." Bandurski explains that over the past decade, China's leadership has identified innovation as the way forward for economic and social development. The promise of innovation has been epitomized by China's tech entrepreneurs, including billionaire founders like Alibaba's Jack Ma and Tencent's Pony Ma. But the dream of innovation has collided with the harsh reality of overwork in a technology sector that seems sapped of opportunities for breakthrough.

Jack Ma and others have advocated a severe culture of overtime work that has become known as "996—working from 9 a.m. to 9 p.m. six days a week.[57] This work culture has been a major reason why lying flat has become so popular with the youth, as young Chinese do not have the same success that they have been promised by billionaires like Jack Ma. The disobedience against the elites by the Chinese youth could translate to disobedience against the government if the Chinese government draws enough ire from their youth.

Bandurski elaborates further that at the technology giant Huawei, this extreme work environment has been dubbed "wolf culture," a climate of fierce internal workplace competition in which workers must either kill or be killed. Observers have drawn a straight line from "996" culture to the "lying flat" movement. He writes that "Lying flat was spawned under the persecution of the 996-overtime culture." When confronted with "996" in 2019, Jack Ma said: "In this world, all of us want to be successful, all of us want a good life, and all of us want to be respected. I ask you, 'how can you achieve the success you want if you don't put in more effort and time than others?'"

Bandurski concludes with "in a society where more open forms of protest, such as labor activism, are quickly suppressed, they have found release, if not relief, in online expression. The 'lying flat' movement, whose forums have drawn upwards of 200,000 members, is one example of this, and a slew of popular online terms have emerged to describe the sense of hopelessness. These include 'leek people' and 'harvesting leeks,' phrases that liken those caught in the struggle of work and consumption to leeks that are constantly harvested under the blade. 'Lying flat-ism,' one Chinese journalist wrote on the Weibo platform, 'is a non-violent movement of non-cooperation by the leek people, and the most silent and helpless of actions.' When one opts out of the cycle, or so the reasoning goes, it is no longer possible to be cut down … (As depicted in a social media illustration) A harvest knife slashes vainly in the air as the plants below fold themselves down toward the

57 Bandurski.

earth. 'Leeks that lie down cannot be harvested so easily,' the caption reads."[58] The Chinese government has underestimated the displeasure of the youth that they will rely on becoming the main driver of the nation's workforce and military force. The "lying flat" movement could make conscription difficult, as China lacks the incentives to motivate the youth of the nation.

"Lying flat" not only shows the cracks in China's economy, but also in the country's potential prospects for a war. Dima Zayaruzny uncovers the causes for the "lying flat" movement. He explains that "lying flat" gained traction as the Chinese economy slowed down after COVID-19 and the end of the "growth miracle" of the late 1990s and 2000s. Young adults were having more problems with finding jobs compared to their parents 20-30 years ago. China had 11.6 million graduates in June of 2023 and a youth unemployment rate of over 20 percent, unwelcome statistics that the government has stopped posting. As Zayaruzny says: "The intense societal pressure to excel academically and professionally has long been a hallmark of Chinese culture, resulting in the youth being shamed for not working hard or underperforming in school. Consequently, young workers are burnt out by this pressure and are unsatisfied with their careers and their professional lives. Concurrently, the Chinese government is implementing policies to divert talent from the private tutoring, property, tech, and finance industries and move them towards sectors it deems essential for the nation's advancement, particularly information technology and advanced manufacturing." Zayaruzny goes further to say that "lying flat" has consequences for China in terms of production from disengaged workers. He cites a Forbes report that said, "disengaged employees are 15 percent less profitable and 18 percent less productive than those who are satisfied with their jobs."

This combination of the "lying flat" movement and unemployment will be detrimental to China's economy. He then writes that "Individuals opting out of high-stress careers can lead to a less efficient allocation

58 Bandurski.

of human capital, potentially resulting in higher unemployment rates and reduced economic output. Moreover, the delayed reopening from COVID-19 restrictions and the ongoing global trade slump have contributed to economic stagnation, posing significant hurdles to recovery." He adds that the worsening relations with America also hurt China's economy as foreign investment in Chinese bonds, companies, and currency decreases. Zayaruzny concludes: "China must prioritize the well-being of its workforce and recognize that workers are the backbone of production and output. Ensuring a balanced work-life environment, offering opportunities for personal growth, and fostering a culture that values individuals beyond their economic contributions are essential."[59] China's faltering economy and dissatisfied youth could halt any further ambitions on the part of the Chinese government.

Given how the West is preoccupied with the Russo-Ukrainian War, China could use the war as a distraction to further its ambitions in the South China Sea. How China can use the Russo-Ukrainian War to their advantage is covered by Bradley Worrell. Worrell, interviewed by Steve Chan, explained that the Soviet Union reaped the benefits of the Korean War, which pitted America and China against each other for 30 years. In much the same way, China may be given some breathing room as America and Russia butt heads over Ukraine. It could take decades for Russo-American relations to normalize. He argues that while China remains America's biggest rival, internal issues such as the housing crisis of 2008 and external issues like the War on Terror have distracted America from the second rising power in the world. Just as Russia sees its breakaway oblasts as separatists, so does China view Taiwan. Much of the world, America included, view Taiwan as a part of China. Chan elaborates on what China and the West have learned from the Russo-Ukrainian conflict by saying "(A) lesson from

59 Zayaruzny, Dima (Bachelors in Economics and Chinese at the University of California, Berkeley), Hoa, Kiet (Editor at the Berkeley Economic Review.) "'Lying Flat': The Demise of the Chinese Workforce and Its Impact on the Chinese Economy." *Berkeley Economic Review*. November 14, 2023.

Ukraine is the importance of unmanned drones. The Chinese have learned that, in a future conflict, unmanned drones will be a very important element. It's much cheaper to produce these drones and to deploy them. Still another lesson is that it will be more difficult for the West to supply Taiwan with armament, because as an island it lacks the easy land access available to Ukraine. Moreover, the Europeans will have less 'skin' in a possible conflict across the Taiwan Strait compared to Ukraine, and they are thus even less likely to involve themselves directly in such a contingency." He adds that it is easier for America to supply Ukraine through Europe, whereas equipping Taiwan during a war would be difficult as the island could be surrounded during the first month of the war. A blockade would not have a military effect, but a psychological effect as people would panic and would run out of gasoline in a matter of months.[60] China could gain both militarily and politically from the Russo-Ukrainian War.

China has enjoyed an era of modernization of its economy and military, which has made the nation second only to the United States of America as the strongest military and economy. This has given the Chinese government leverage to push their agenda of owning not only Taiwan, but the South China Sea. This has the consequence of deteriorating relations with their neighbors, as Chinese naval vessels keep intruding on the EEZs of other nations and stealing fish and other resources. Internal policies of China have become a problem for Chinese youth and minorities mainly in Tibet and Xinjiang. The former will cause a problem with military recruitment, while the latter will be an issue if the people of Tibet and Xinjiang suspect weakness in China and rebel, which will cause the diversion of military units and supplies that will be pivotal to the invasion of Taiwan.

---

60 Worrell, Bradley (Writer and editor for the *Colorado Arts and Sciences Magazine*). "Is today's Ukraine tomorrow's Taiwan?" *Colorado Arts and Sciences Magazine*. Published by the University of Colorado in Boulder, Colorado. February 10, 2023. https://www.colorado.edu/asmagazine/2023/02/10/todays-ukraine-tomorrows-taiwan.

Chapter 3: Taiwan and a Nation's Need to Survive

Ever since its inception as an independent entity, Taiwan has been threatened by conquest from the Chinese mainland. For 60 years, Taiwan had to rely on the military supplies from America to arm up its defense against invasion. The support from the United States is due to US government efforts to dissuade the Chinese government from aggressive expansion. This caused a further rift between China and the United States as they vie to determine the fate of Taiwan, which could determine the future of the South China Sea. Taiwan will have to choose how to navigate through the tenuous times and prepare for the worst-case scenario of being a frontline nation in a major war with China.

While China has historical and geopolitical reasons to not let the Taiwan issue go, Taiwan also finds it difficult to have a normal relationship with China. This is explored by Lung Chu Chen in his 2016 book. Chen brings up that during the presidency of Ma Ying Jeou (2008–2016), Taiwan and China signed the Cross-Strait Service Trade Agreement (CSSTA) in June of 2013. In the CSSTA, Taiwan and China agreed to allow citizens of both countries to do exchanges in banking, healthcare, transportation, travel, communications, and construction, and to permit cross-strait investments. Jeou was criticized for leaving Taiwan undefended against a country that wishes to absorb them. When the CSSTA was going through Taiwanese legislation, hundreds of protestors stormed the Yuan legislative building in Taipei, Taiwan's capital, on March 18, 2014. Five days later, Ma doubled down on his support for CSSTA and urged for its passage. This led to hundreds of thousands of Taiwanese protestors taking to the streets of Taipei in support of those who occupied the Yuan legislative building. The occupation ended on April 10 after KMT Legislative Speaker Wang Jin-pyng pledged to enact legislation to monitor cross-strait agreements before looking at the CSSTA. Chen concludes by writing that not only did this protest change how Taiwanese legislation looked at cross-strait agreements, but it also demonstrated that the younger generations of Taiwanese people were becoming more invested in their country's politics.[61] What the reaction to the CSSTA shows is that internal disagreements can make it difficult for the Taiwanese to engage in diplomacy with China.

With the threat of war being ever-present, Taiwan has to deal with talks about conscription. John Dotson discussed in his journal article how in March 2022, Defense Minister Chiu Kuo-cheng, Interior Minister Hsu Kuo-yung and other ROC officials met to talk about increasing service time for their soldiers. Ever since then, the Taiwanese have talked about expanding the conscription service. That raises questions about how to adequately train these new units, and about the potential financial costs.

Conscription is a big enough issue for Taiwan that then President Tsai Ing-wen made a speech on December 27, 2022, about the "Military Force Realignment Plan." She said that "'Taiwan stands on the frontlines of authoritarian expansion, at the vanguard of the global defense of democracy. Only by preparing for war can we avoid it … Only by strengthening our self-defense capabilities can Taiwan more effectively protect our national security and interests while garnering further international support." This plan, which went into effect in 2024, applies to all able-bodied males born on or after January 1, 2005.

---

61 Chen, Lung-chu (Scholar and Professor of Law at New York Law School, specializing in international law, human rights, the United Nations, and Taiwan). *The U.S.-Taiwan-China Relationship in International Law and Policy.*: The U.S.-Taiwan-China Relationship in International Law and Policy - Lung-chu Chen - Google Books. Published by Oxford University Press. New York, New York, U.S.A. 2016. 33.

The "Military Force Realignment Plan" will also divide military actions into four categories:

- "Main Battle Troops": active-duty military volunteers that will go from 180,000 men to 210,000 men and will be the frontline forces.
- "Garrison Troops": mainly conscripts who will focus primarily on infrastructure protection and territorial defense.
- "Civil Defense System" or "Alternative service personnel": used for disaster relief, medical services, and military support operations.
- And a "Reserve System": of a million reservists meant to "replenish our main battle force with retired volunteer soldiers, and our garrison force with former mandatory servicemembers."

Tsai mentioned the need for the Taiwanese army to train in new weaponry such as "Stinger missiles, Javelin missiles, Kestrel rockets, drones, and other new types of weapons (which) will be added in accordance with mission requirements to meet the needs of modern warfare."[62] While Taiwan would have to deal with an overwhelming force in the form of the Chinese military, the modernization of the Taiwanese military and reformation of conscription helps Taiwan's chances to withstand an assault.

However, Taiwan's ability to conscript is hindered by the low birthrate in the country. The reason for said birthrate is explained by George Carter, who writes that Taiwan's fertility rate is 1.07, the lowest birthrate in the world. He attributes this to low marriage rates in Taiwan due to economic burdens, attitudes towards childbearing, and higher

62 Dotson, John (deputy director of the Global Taiwan Institute and associate editor of the *Global Taiwan Brief*). "Taiwan's 'Military Force Restructuring Plan' and the Extension of Conscripted Military Service.": https://globaltaiwan.org/2023/02/taiwan-military-force-restructuring-plan-and-the-extension-of-conscripted-military-service/. *Global Taiwan Brief* Vol 8. Issue 3. Published by the Global Taiwan Institute in Washington, D.C., U.S.A. February 8th, 2023. 2-3.

participation in the workforce by women. The low birthrate means that there will be a shortage of men to fill military positions. Carter claims that Taiwan's birthrate problems can be traced back to 1958, when the country turned from an agricultural economy into a manufacturing powerhouse. The large family that was needed to tend to the farm is now unnecessary. He adds that in 1950, Taiwan contracted American demographer George Barclay to curb a rising population. His solution was to expand sex education and the availability of contraceptives. Taiwan also expanded the availability of abortion, which has led to 500,000 annual abortions. Carter's solutions to this problem are cancelling all volunteer programs and replacing them with a more rigorous conscription program, loosening immigration laws, using new technologies like automation, and extending the service age.

Carter explains that here are two approaches to military structure. The first is the All-Volunteer Force (AVF), where volunteers sign a contract to serve a certain amount of time. The second approach is conscription, where all able-bodied people in a country would serve for varying lengths of time. He writes the AVF program failed in Taiwan due to poor execution and fiscal challenges. While South Korea has a conscription time-range of 18 to 23 months, and Israel has a conscription time range of 24 to 36 months, Taiwan only has four-month conscription time. Carter warns Taiwan could repeat the same mistake the Ukrainians made in 2013, when Ukraine underestimated outside threats, switched from a conscription service to an AVF program, and lost Crimea and Donbass as a result. In 2014, Ukraine reinstated conscription less than a year after its termination. Carter also suggests that migration can partly offset a low birthrate. In 2021, America had a net migration rate of 3.03 per 1,000 people while Taiwan has a net of 0.75 per 1,000 people. Taiwan has limited paths to citizenship and naturalization. In contrast, America has multiple pathways to citizenship and naturalization.

Carter recommends Taiwan eliminate the AVF and replace it with a professional military. He writes, "Taiwan should reconsider their immigration policies. Taiwan should also consider creating a mercenary force made up of foreign nationalities like the French Foreign Legion. The failure of the AVF to be properly manned shows why it should be terminated." Carter adds that the AVF is too costly for Taiwan (about 47% of Taiwan's defense budget), especially at the expense of American arms and readiness. He recommends that Taiwan reinstate conscription and recruit volunteers to create an effective fighting force and reserves like Israel. The knowledge, skills, and abilities of a professional soldier are developed over years not months. As of 2020, 3,032,774 men were eligible for conscription. By 2065, this number is projected to go down to 424,000. Due to the aging population, Carter suggests the age restriction can be raised from 45 to 50 years old, giving Taiwan much needed manpower. Taiwan may also benefit from the unmanned surface vessels that the USN is experimenting with. The model of the French Foreign Legion is one in which people could gain French citizenship after 3 years of service. Carter believes that a Taiwanese Foreign Legion will serve well to help the manpower shortage.[63] To summarize:

1) Taiwan had a population problem, but they over-corrected it.
2) This led to a smaller pool of able-bodied conscripts for their military.
3) The terms of conscription have not been effective.
4) Limited paths to citizenship and naturalization also hinder population growth.

Taiwan's lack of manpower will not only need to be backed up by firepower, but also by cooperation among Taiwan and partnering nations that wish to curb Chinese aggression.

63 Carter, George (Author). "Projecting Taiwan's Military Manpower Supply, 2020-2065.":Title page for etd-1005121-004121 (nsysu.edu.tw). Global Human Resource Management English MBA Program. Published by National Sun Yat-sen University. Kaohsiung, Taiwan. November 2021. 6, 12, 22-23, 36-37, 85-88, 91-92, 100.

While China is preparing for war with Taiwan now, in the past it did propose a program to unite peacefully, called the One Country Two Systems program (1C2S) Richard C. Bush writes that the origin of 1C2S was in 1990, when China implemented "The Basic Law" for Hong Kong, which acts as the basis for the program. The 1C2S proposal would make Taiwan a special administration region of China. Taiwan would no longer exist as a separate entity, and the Chinese flag would fly over the island. Taiwan would be politically and administratively subordinate to China, which would control Taiwan's foreign and defense affairs. Taiwan would no longer be a platform for America. However, Bush added that China claimed that Taiwan's economic and social life would continue as before. Taiwan would have a "high degree of autonomy" and individuals in Taiwan would administer their nation. Taiwan would enjoy protection of civil and political rights, and an independent legal system.

However, when Hong Kong protested in 2019, China's response proved that China would use force to maintain its hold on all claimants, Taiwan included. Bush wrote that there are reasons for Taiwan not to accept the 1C2S proposal. For one, American support for Taiwan's security has continued since the Reagan Administration. The end of the Soviet Union made China more isolated and America more determined to help Taiwan. The fall of the Soviet Union also opened an advanced weapons market, which Taiwan took advantage of. However, the main reason China has not reunified with Taiwan, is that the Taiwanese people do not support the 1C2S proposal. A 2019 poll by the Election Study Center of National Chengchi University found that 79 percent of Taiwanese people reject 1C2S, with only 10 percent supporting it.[64] This creates a greater picture as to the divide between China and Taiwan, which remains the heart of the South China Sea issue.

---

64 Bush, Richard C. (American expert on China affairs. Since 2002, he has served as the director of Center for Northeast Asian Policy Studies [CNAPS] of the Brookings Institution, and concurrently as the inaugural Michael H. Armacost Chair in Foreign Policy Studies.). *Difficult Choices: Taiwan's Quest for Security and the Good Life*. :Difficult Choices: Taiwan's Quest for Security and the Good Life

While American supplies to Taiwan are indispensable, their relationship and any plans to strengthen said relations could spark a war between America and China. Taiwan's relations with America and their regional neighbors are covered by Denny Roy in his journal article. Roy explains America's position by citing former Secretary of State Mike Pompeo, who said at the Indo-Pacific Business Forum in July 2018 that the Trump Administration's goal in the region is "to [create] a regional order, [where] independent nations … can defend their people and compete fairly in the international marketplace." Roy explains that Taiwan's Pacific Islands Policy and America's Free and Open Indo-Pacific strategy (FOIP) share a common interest of deepening connections with regional powers to counter Chinese ambitions. However, closer ties between America and Taiwan could provoke China to lash out. It would be sensible for America to quietly support Taiwan and improve their standing with Pacific nations. But Taiwan's options for regional partners are slim, as only six of the 17 Pacific nations recognize Taiwan. Those six nations are Kiribati, Marshall Islands, Nauru, Palau, Solomon Islands, and Tuvalu. Worse yet, China is trying to buy off Taiwan's allies. Tsai Ing-wen said at the 2018 national day that "The best way to defend Taiwan is to make it indispensable and irreplaceable to the world." Both President Tsai and Foreign Minister Joseph Wu "have argued that Taiwan is relevant to FOIP in three ways:

1) As a country that shares 'values and beliefs' with the United States,
2) Through its ability to help build capacity in regional countries friendly to the United States, and
3) As a landmass that occupies an 'irreplaceable' and 'important geostrategic position.'

- Richard C. Bush - Google Books. Published by the Brookings Institution Press. Wahington D.C., U.S.A. 2021. 126-129.

Roy adds that "Washington sees Taiwan in two ways. First, it shares liberal democratic values with the United States … Second, US strategists still see Taiwan largely as a geopolitical flashpoint for potential US-China conflict. Mike Pence said that Taiwan's democratic system was "a better path for all the Chinese people." Taiwan stands as a flashpoint for American and Chinese tensions.[65] America and Taiwan share similar visions for the Indo-Pacific, but the threat of a war with China could hamper further cooperation between the two nations.

Much like all nations in the Indo-Pacific, Taiwan faces internal problems when it comes to land claims. This political issue is examined by William Yang, who cites examples of land disputes that occurred in 1956, when Taiwan took control of the 46-hectare (114-acre) islet Itu Aba, or the Taiping Island of the Spratly Islands. Since then, China, Vietnam, and the Philippines have all laid claim to the island. Yang said in 2024 that "It (Taiping Island) hosts 200 members of Taiwan's coast guard and contains a hospital, a farm and a runway to facilitate military resupply missions. In January, Taiwan completed a dredging project that would allow larger vessels to dock at the wharf." However, in 2016, Taiping Island was declared a reef by a Hague tribunal under UNCLOS. This means that Taiping Island cannot be under a claim of sovereignty. Yang explained that the result of this led to the Kuomintang Party to criticize President Tsai for not enforcing Taiwan's claims on the island. Yang quotes a Kuomintang leader: "'Since the South China Sea arbitration downgraded Taiping Island to a reef in 2016, the Tsai administration hasn't protested the ruling,' Ma Wen-chun, a Kuomintang legislator, wrote on her official Facebook page." Yang suggests that Tsai visiting Taiping Island would strain relations between Taiwan and China and the Philippines. He cites Ivan Kanapathy, who

---

65 Roy, Denny (Senior Fellow and Supervisor of the POSCO Fellowship Program at the East-West Center in Honolulu). "Taiwan's Potential Role In The Free And Open Indo-Pacific Strategy Convergence in the South Pacific." :Taiwan's Potential Role in the Free and Open Indo-Pacific Strategy (nbr.org), NBR Special Report #77. Published by the National Bureau of Asian Research. Seattle, Washington, U.S.A. March 2019.

served on US President Donald Trump's National Security Council as director for China, Taiwan and Mongolia, who said in part: "A trip to Taiping Island would be counterproductive to Tsai's New Southbound Policy and the diplomatic progress that Taipei has made with regional partners over the past 8 years." This is due to the overlapping claims Taiwan and other southeast Asian nations have over Taiping Island.

Su Tzu-yun, a military analyst at the Taipei-based Institute for National Defense and Security Research, told VOA by phone that "Since tensions remain high between China and the Philippines and Vietnam in the South China Sea, it'll be favorable for Taiwan if the government uses more peaceful ways to safeguard its territorial integrity."[66] Though Taiwan wants to have stronger ties to its neighbors, this could come at the cost of their claims on land such as Taiping Island.

Despite being a conflict from the other side of the world, Taiwan and its people look at the Russo-Ukrainian War and take note on what could happen if Taiwan and China were to go to war. The Russo-Ukrainian War, from a Taiwanese Perspective, is explored by Russell Hsiao and others in the June 2023 edition of *Global Taiwan Brief*. The authors explain that the invasion of Ukraine resonated with the Taiwanese public, to the point that the media was parroting the phrase "Ukraine today, Taiwan tomorrow" for the first months of the Russo-Ukrainian War. Other Asiatic nations have taken notice, as when Japanese Prime Minister Fumio Kishida visited Ukraine in March 2023. He said, "Ukraine might be East Asia tomorrow … We must show there are consequences to the attack."

The *Global Taiwan Brief* authors go on to add that Taiwan has given more than $113.6 million in financial support to Ukraine and provided more than 700 metric tons in humanitarian supplies. However, given Ukraine's reluctance to enrage China, Taiwan has

---

66 Yang, William (an East Asia Correspondent, with his work appearing in various publications including VOA now Stantec). "Taiwan Faces Sovereignty Dilemma in South China Sea Amid Chinese Pressure." Voice of America. March 22, 2024.

had to send these supplies through city governments, churches and regional NGOs, or through Eastern European countries. The authors wrote that "Despite these limitations, Taiwan is still in the top five Indo-Pacific countries in terms of bilateral aid sent to Ukraine, trailing only Japan, Australia, and South Korea, while leading New Zealand. In the East and Southeast Asia regions, Taiwan is among the leading countries, coming in as the third-largest donor of humanitarian aid to Ukraine. Very little aid to Ukraine has come from nations in the latter region, except for Thailand, Singapore, and Timor-Leste." This fervor in support is explained by former Taiwanese ambassador to the United States Hsiao Bi-khim, who said at a forum in Washington, D.C., in May 2023 that "Ukraine's survival is Taiwan's survival. Ukraine's success is Taiwan's success. Our futures are closely linked." Taiwan's top diplomat went on to say, "I think pushing back on aggression is the key message that will help to deter any consideration or miscalculation that an invasion can be conducted unpunished, without costs, in a rapid way. We must ensure that anyone contemplating the possibility of an invasion understands that, and that is why Ukraine's success in defending against aggression is so important also for Taiwan."

The authors consider that America is supporting Taiwan as they write: "The United States has likewise remained committed to providing Taiwan with the tools necessary to defend itself. The fiscal year 2023 National Defense Authorization Act (NDAA) authorized an additional USD $10 billion for Taiwan. In March 2023, the State Department approved a USD $619 million sale of new weapons to Taipei. Meanwhile, regular US military transits in the Taiwan Strait have continued apace, despite China's provocations." The authors then cite a poll by the Taiwan Public Opinion Foundation (TPOF), which shows that nearly 60 percent of Taiwanese correspondents think that their country is following the same footsteps as Ukraine. While polls show that only 29.6 percent of Taiwanese people think they can win a war against China, 60-70 percent are willing to fight for the nation's sovereignty.[67] The outcome of the Russo-Ukrainian War could shape how Taiwan and China will react to each other and their neighbors in the South China Sea.

However, because the Russo-Ukrainian War is so recent, its impact on the Sino-Taiwan situation remains uncertain. Erin Hale writes that while observers wonder if China will attack Taiwan just as Russia attacked Ukraine, it is unlikely that China will attack Taiwan anytime soon. CIA Director William Burns said that although Xi was "unsettled" by Russia's setbacks in Ukraine, he stands on the idea that China will be ready to invade Taiwan by 2027. Burns added that "We know as a matter of intelligence that he's (Xi) instructed the People's Liberation Army to be ready by 2027 to conduct a successful invasion," and "Now, that does not mean that he's (Xi) decided to conduct an invasion in 2027, or any other year, but it's a reminder of the seriousness of his focus and his ambition."

Tai Ming Cheung, director of the Institute on Global Conflict and Cooperation at the University of California, San Diego, told Al Jazeera in 2023 that there is a difference between the leaders of Russia and China: "One of the assessments is that Putin acted very impulsively because of his imperial ambitions, and he has his own worldview," Cheung said. "Xi, I think, he's much more pragmatic. He's very cautious. I don't think he's hot-headed enough and he's not a risk taker, and an invasion against Taiwan is one of the biggest acts that he would do. It's a very high-wire act and the chances of success are not clear."

Hale explains that, to China, The Russo-Ukrainian War contrasts with the Gulf War and Iraq War as the stalemate of the former is different from the decisiveness of the latter two. Despite this, Hale argues that there are benefits China can reap from the Russo-Ukrainian War. Russia's hypersonic missiles, used against Odessa,

---

67 Hsiao, Russell (the executive director of the Global Taiwan Institute (GTI) and editor-in-chief of the *Global Taiwan Brief*), Dotson, John (Associate Editor of the *Global Taiwan Brief*), Reid Marshall (Staff Editor of the *Global Taiwan Brief*). *Global Taiwan Brief* Vol 8. Issue 13. Published by the Global Taiwan Institute in Washington, D.C. June 28, 2023. 1-2, 8, 13.

are something that China wants to invest in. Another lesson China took from the Russo-Ukrainian War is the use of nuclear blackmail, which has prevented Western nations from enacting a no-fly zone in Russia. With the second largest economy in the world, China could stack economic blackmail on top of nuclear blackmail in the time of war against Taiwan. Hale adds that Taiwan has also been taking notes and has been investing in portable arms such as Stinger surface-to-air missiles, Javelin anti-tank missiles, and Harpoon anti-ship missiles. The island's leaders see that FM radio to commercial satellites can be used to inform citizens and to gain global support.[68] While both China and Taiwan are taking notes, the Russo-Ukrainian War will be an ever-evolving issue until the conflict's conclusion.

The Russo-Ukrainian War has given Taiwan ideas of how a smaller army could withstand the force of a larger one. Dereck Grossman reports that in May of 2022, Taiwanese Foreign Minister Joseph Wu told CNN's Fareed Zakaria that "We try to see what we can learn from Ukraine in defending ourselves." Wu noted the effectiveness of "small personal weapons to go against a large enemy.

"All of the males (in Ukraine) are having the determination to defend the country," he added. "They want to serve in the military. They want to go to the war zones to fight against Russia. That kind of spirit is enviable for the Taiwanese people." In April 2022, Taiwan enacted a new Defense Mobilization Agency, which is part of the Ministry of National Defense (MND). This agency issued a "war survival guide," which Grossman said "explained various crisis responses, like how to find bomb shelters as well as food and water via smartphones, and how to prepare first-aid kits. But the agency is not charged with organizing civilian resistance, nor does the handbook provide guidance on it. It will be important to see if Ukraine moves Taiwan toward further civil-military integration and synchronization."

---

68 Hale, Erin (a journalist working at Al Jazeera). "Will Russia's struggle in Ukraine help Taiwan—or hurt it?" Al Jazeera. February 23, 2023. https://www.aljazeera.com/features/2023/2/23/will-russias-ukraine-struggle-help-taiwan-or-hurt.

Grossman mentions, however, that while Taiwan invests in anti-ship cruise missiles, sea mines, and surface-to-air missiles, they also buy bigger items that can detract from the cheaper and more effective weapons. For example, Taiwan bought 66 f-16V fighter jets from America. Taiwan's new strategy sees them fighting the People's Liberation Army (PLA) on the shores of Taiwan rather than on the Taiwan Strait. Grossman adds that both America and Taiwan have their concerns when it comes to the preparations for a war. He writes: "Americans have been skeptical and concerned over Taiwan's level of resolve or will to fight. To be sure, since transitioning from conscription to an all-volunteer force, Taiwan's military has struggled mightily to recruit young people ... A telephone survey conducted by the Taiwanese Public Opinion Foundation, a private think tank in Taiwan, of about 1,000 Taiwanese aged 20 or older from April 17 through April 19 indicated that 53.8 percent of respondents somewhat or completely disbelieve that the United States would come to Taiwan's defense, up from 28.5 percent last October. In other words, America's reluctance to directly intervene on Ukraine's behalf against Russia is fueling a sense in Taiwan that Washington might not be as reliable as previously thought."[69] Though Taiwan has gained insight into how to fight a larger foe, the Taiwanese people are questioning America's ability to respond to threats.

Taiwan has looked for other nations to help them in their defense from China. Australia has been one nation Taiwan has looked to for support, as described by Andrew Greene and Stan Grant. They bring up that Taiwanese Foreign Minister Joseph Wu warned that China is preparing for war, and that Wu called on Australia to "increase intelligence sharing and security cooperation as Beijing intensifies a campaign of military intimidation." On ABC's *China Tonight* program,

---

69 Grossman, Dereck (a senior defense analyst at the nonprofit, nonpartisan RAND Corporation. He formerly served as an intelligence adviser at the Pentagon.) "Ukraine War Is Motivating Taiwan to Better Secure Its Own Future." RAND Corporation. May 13, 2022. https://www.rand.org/pubs/commentary/2022/05/ukraine-war-is-motivating-taiwan-to-better-secure-its.html.

Wu said Taiwan would be ready to repel an attack from China. Wu said, "The defense of Taiwan is in our own hands, and we are absolutely committed to that ... If China is going to launch a war against Taiwan we will fight to the end, and that is our commitment ... I'm sure that if China is going to launch an attack against Taiwan, I think they are going to suffer tremendously as well." Wu said Taiwan seeks better relations with Australia as they are like-minded nations. He added, "We would like to engage in security or intelligence exchanges with other like-minded partners, Australia included, so Taiwan is better prepared to deal with the war situation ... And, so far, our relations with Australia [are] very good and that is what we appreciate."

Greene and Grant explain however, that Australia does not recognize Taiwan and calls for a peaceful resolution between China and Taiwan. Wu thanked Australia for their support in entering the Trans-Pacific Partnership trade pact, now known as the CPTPP. He said the Taiwanese appreciate the supplies given to them by Australia. They are also grateful for the activities of the QUAD and the partnership of Australia, the United Kingdom and the United States (AUKUS). Wu mentioned that unlike Australia, which recently completed a deal to acquire nuclear submarines, Taiwan needs to emphasize asymmetric warfare. He concluded by talking about how Taiwan's defense is about preparing for invasion, but if China chooses to blockade Taiwan, then "Taiwan doesn't have a Plan B."[70] This push for strengthening ties with Australia is a way for Taiwan to create a better deterrence against the Chinese government.

Taiwan has been a miracle both in the economic sense and as a nation. Not only has the island nation managed to create a modern

---

70 Greene, Andrew (ABC's defence correspondent and has covered federal parliament since 2004), Grant, Stan (Vice Chancellor's Chair of Australian/Indigenous Belonging at Charles Sturt University. He was formerly ABC's Global Affairs and Indigenous Affairs Analyst.) "Taiwanese Foreign Minister warns his country is preparing for war with China, asks Australia for help." ABC. October 4, 2021. https://www.abc.net.au/news/2021-10-04/taiwan-preparing-for-war-with-china/100511294.

economy, but it has survived as a country despite China's saber-rattling. Through its history of defiance against a totalitarian state, Taiwan has become a beacon of democracy in the South China Sea. Taiwan's continued existence gives the major powers of the South China Sea the inspiration to support the nation's attempt to modernize the army against China. If war does break out between China and Taiwan, then the island nation must stand strong to be a rallying point for the fight against authoritarianism.

Chapter 4: America: The Titan Staring at a Rising God

Though America does not want to fight a war with China, it has become a difficult scenario for the American government to escape. Nong Hong of the Institute for China-America Studies explained this by showing the different approaches to China taken by the Obama Administration and the Trump Administration. During a trip to China to discuss matters of American military vessels near Chinese waters, Obama met with Xi to create a "base of cooperation and coordination rather than conflict." The meeting resulted in both countries agreeing to notify each other of military activities. However, Nong notes a split in the Trump Administration as to how to deal with an aggressive China. Former Secretary of Defense James Mattis on a trip to Tokyo said that there is "no need for dramatic US military moves in (the) South China Sea." This contrasts with then senior counselor Steve Bannon saying, "there is no doubt (that the United States is) going to war in the South China Sea in 5 to 10 years." Sean Spicer, Trump's White House press secretary, said "we're going to make sure that we defend international territories from being taken over by one country." On February 12, 2017, *The Navy Times* wrote that "the [American] Navy is planning fresh challenges to China's claims in the South China Sea." Hong recommends that China and the United States negotiate on their military relations and their legal views on maritime issues. They could also support one another on guarding against misjudgment, reducing confrontation, and managing and controlling crises.[71] Though as of this writing, America is trying to keep a diplomatic approach to China's ambitions, only time will tell if this approach works.

While the United States and China are currently in a standoff, it is important to look to their histories to find a common solution that could stand the test of time. Such a solution could be found in a 2016 book by John Pomfret. Pomfret explains that the history between America and China began in the 1800s, when American traders did business with the Chinese. The exchange between the two resulted in America getting Chinese art, food and philosophy, and China receiving American science, educational theory, and technology. Pomfret claims that further on in their history, American missionaries were instrumental in Chinese modernization. Missionaries introduced Western science, critical thinking, sports, industry and law to China, and established the first hospitals and universities. American women missionaries fought against female infanticide and foot-binding, helping China accomplish its greatest human rights advances.[72] What can be taken from Sino-American history is that the exchange of ideas from both nations has created a respect between the two. Said respect must be remembered for China and America to maintain peaceful terms with each other. Both governments should make the point that their trade of goods and ideas make them indispensable to one another.

Though the United States has a significant presence in the South Pacific, America will need its partners in the region to give support in the event of war. Dimitri Zabelin elaborates on how America could

71 Hong, Nong (Executive Director Senior Fellow, Institute for China-America Studies). "Understanding the Freedom of Navigation Doctrine and China-US Relations in the South China Sea: Legal Concepts, Practice, and Policy Implication.": Understanding-FONOP-Doctrine-Report-FINAL.pdf (chinaus-icas.org). Published by Institute for China-America Studies in 2017. 19, 22-23.

72 Pomfret, John (American Journalist and Writer). *The Beautiful Country and the Middle Kingdom: America and China, 1776 to the Present.* :The Beautiful Country and the Middle Kingdom: America and China, 1776 to the ... - John Pomfret - Google Books. Published by Jenry Holt and Company, New York, New York, U.S.A. 2016. 2-3.

improve relations with its partners in the Pacific. Zabelin writes that Trump's trade war left China more insular and led to fewer countries backing the United States. In 2020, China and other Asia-Pacific nations signed the Regional Comprehensive Economic Partnership (RCEP). The partnership is the world's largest trade agreement, with the 10 nations of ASEAN, Australia, Japan, South Korea and New Zealand taking part along with China. However, whatever goodwill the Chinese engendered from RCEP was dashed with China's lack of cooperation and transparency when it came to the COVID-19 outbreak. It got to the point that countries other than the United States, including Australia, have called for an investigation into China's involvement in the COVID-19 outbreak.[73] America can use the failures of the Chinese government in the case of COVID-19 to further unite its partners on the mission of curbing Chinese aggression.

The United States of America took note in 2013, the year China ramped up its push into the South China Sea. Sheldon W. Simon cites a March 11, 2013, foreign policy address by then American National Security Advisor Tom Donilon, in which Donilon told the Asia Society: "The United States is implementing a comprehensive multidimensional strategy: strengthening alliances; deepening partnerships with emerging powers; building a constructive relationship with China; empowering regional institutions; and helping to build a regional economic architecture that can sustain shared prosperity." He added that the American action "is also a response to the strong demand signal from leaders and publics across the region for US leadership, economic engagement, and sustained attention to regional institutions and defense of international rules and norms."

Simon also cited then commander of the United States Pacific

---

73 Zabelin, Dimitri (Associate at LSE IDEAS, having studied political economy at the University of California, Berkeley for both his undergraduate and masters). "How Can America Challenge China's Political Ambitions in an Age of Deglobalisation?.":How can America challenge China's political ambitions in an age of deglobalisation? - *LSE Research Online*, LSE IDEAS Strategic Update September 2021. Published by LSE Ideas. London, Britain. September 2021. 6, 8-10.

command Admiral Samuel J. Locklear who explained to the United States–Indonesia Society in Jakarta on February 8, 2013, the reasons that America re-shifted its attention on the Pacific: "The many nations who associate themselves here include five of our nation's seven treaty allies, the largest economies of the world. ... The most populous nations of the world are also represented, which also include the first, second, and third largest democracies. ... Nine of the world's largest ports are in the Asia-Pacific. The sea lanes here are the busiest in the world through which pass over half the entire world's container cargo every day and over 70 percent of ship-based energy. ... The Asia-Pacific is also the most militarized area in the world with seven of the world's ten largest standing militaries, the world's largest and most sophisticated navies, and five of the world's declared nuclear states."

That same year, America identified six strategic partners in the Pacific: Singapore, the Philippines, Thailand, Indonesia, Malaysia, and Vietnam. Simon points out Thailand especially as "US and Thai armed forces hold over 40 combined military exercises per year, and tens of thousands of Thai military officers have taken part in US armed forces training and educational programs." A year earlier, the Pacific Command's order of battle wrote that by 2020 60 percent of America's Navy would be stationed in the Pacific. This includes F-22 and F-35 planes, Virginia class submarines, Global Hawk and MQ-9 unmanned aerial vehicles and U-2 aircraft.[74] America wants to be sure it can counter any move China makes.

Though the United States is thousands of miles away from the South China Sea, America does have bases that can cause trouble for China in the case they go to war. One of these bases is explored by Jiacheng Li in his journal article. The base in question is Diego

---

74 Simon, Sheldon W. (1937-2021) (Sheldon Weiss Simon, Professor Emeritus of Political Science at Arizona State University, who passed away on January 2, 2021)."The US Rebalance and Southeast Asia." :The US Rebalance and Southeast Asia on JSTOR. *Asian Survey* Vol. 55, No. 3 (May/June 2015), pp. 572-595. Published by University of California Press, Oakland, California, U.S.A. 2015. 474-475, 577, 579.

Garcia, which America received from Britian in 1966 in exchange for their "North Star" nuclear missiles and launch systems. As one of the biggest strongholds in the Indian Ocean and located 2,000 km west of the Malacca Strait, Diego Garcia could become a thorn in the side of China in the event of a war, according to Li. The military base is capable of forward operations, supportive supplies and strategic projections. It can meet the operational needs of US carrier battle groups, nuclear submarine forces, amphibious fleets, strategic bombers, and large transport aircraft. That is the reason why Diego Garcia has garnered the name the "unsinkable carrier on the Indian Ocean." The base also played a role in the Gulf War, Afghanistan War, and Iraq War. It helps matters that India has established the East Naval Command in the Nicobar and Andaman Islands to exert control over the area west of the Malacca Strait. Since the George W. Bush Administration, America has not only regarded India as a partner in counterterrorism, but also a partner to constrain China's growing influence. Said partnership exists to this date through the security arrangements called QUAD, an organization that includes America, India, Australia, and Japan.[75] Bases like Diego Garcia and partners like India will prove useful to close the geographic gap between America and the South China Sea.

America has benefited from its presence in the South China Sea since World War 2 and is expected to remain active in the region. The United States not only enjoys good economic relations with the nations in the region, but it also has military relations with South Korea, Japan and the Philippines. The latter two have treaties with America that if they were attacked, the United States must defend them. If anything, America would face economic and geopolitical fallout if they left the region and gave China a free hand. The aggression of the Chinese government has caused America to defend its interests in the region.

---

75 Li, Jiacheng (an associate professor at Liaoning University's School of International Studies in Liaoning, China). "China's Indian Ocean Strategy: Developing China's Indian Ocean Strategy: Rationale and Prospects." *China Quarterly of International Strategic Studies* Vol. 03, No. 04, pp. 481-497 (2017). Published by World Scientific Publishing Co Pte Ltd. Singapore. 485-487.

Americans will continue to develop stronger ties with their South China Sea partners to secure their investments in the region and contain the ambition of their biggest rival.

Chapter 5: Japan: The Rise of a Past Military Power

Since the end of World War 2, Japan has been staying away from its militaristic past and has focused on the peaceful side of its culture. This is largely due to the United States's pacification of Japan after World War 2. Japan, nonetheless, rebuilt after the devastation of the war and managed to become one of the biggest economic powers in the world. In recent times, however, China has passed Japan as the biggest economy in Asia, and the relationship between China and Japan has become tenuous ever since. This is not helped by the ambitions of the Chinese government to take over the South China Sea, which is not only where Japan exports most of its goods, but also its main import channel for food and energy resources.

While China and America remain the predominant powers in Southeast Asia, Japan is still a major player in the region. Paul Midford explains that Japan has one of the biggest economies in the world and is one of the greatest maritime powers in the region. Midford elaborates that since the turn of the 21st century, Japan has developed two new policy instruments: the Japanese Coast Guard (JCG) and more recently the Maritime Self Defense Force (MSDF). Since 2010, Japan has been forming bilateral initiatives to counter China's rising power. Due to pre-existing restrictions such as Article 9 of the Japanese Constitution, which states that Japan can only have a self-defense force and not fight an offensive war, Japan will not fight a direct war in the foreseeable future.

Since the 1970s, Japan has used Official Development Assistance (ODA) to help stabilize the South China Sea and the Malacca Strait. Japan has assisted Indonesia, the Philippines and Vietnam (all of which are members of ASEAN, the biggest geographic partnership of Southeast Asia) to create economic development. The total aid to these three countries from 2008 to 2012 reached nearly $49 billion. Japan participated in anti-piracy exercises with Malaysia and Thailand in 2007. Japan has been engaging with ASEAN on a multilateral level since 1977. In late 2004, a People's Liberation Army (PLA) newspaper accused Japan of "using the fight against piracy as an excuse to cover Malacca with its (Japan's) military attack power." In 2010, Japanese Foreign Minister Katsuya Okada claimed that "Japan cannot be indifferent … the South China Sea is … a strategic point for transportation … Consequently, if it is extremely unstable, then it will impede commerce to Japan." In September 2010, a Chinese fishing boat allegedly rammed into two Japanese Coast Guard (JCG) vessels, leading to the arrest and prosecution of the captain of the Chinese fishing boat. This incident led the growing Japanese perception of China as a military threat.

Japan has communicated with ASEAN about China's ambitions in the South China Sea. Chinese Vice Foreign Minister Cui Tiankai said, "Japan should think comparatively what will truly serves Japan's interests." A Chinese journalist with ties with the Chinese foreign policy translated Tiankai's quote that "Japan would find it annoying if China got involved in a row between Japan and South Korea over the issue of the Takeshima islets and backed Seoul at an international meeting … Parties not in the dispute should remain quiet." In late September 2011, Japanese Prime Minister Yoshihiko Noda and Filipino President Benigno Aquino "concluded a strategic partnership." This led to more cooperation between both countries' navies and coast guards. The Noda Administration (which governed from September 2011 to December 2012) announced that it would use ODA to fund an upgrade for the Philippines Coast Guard. Japan committed to a transfer of two 40-meter 1,000-ton-class large patrol ships and 10 180-ton-class patrol ships. However, the Japanese Coast Guard itself does not carry any weapons. What they offer other nations is the surveillance, thus denying China the ability to use stealth or surprise to move on their territorial claims.

Japan's 2010 National Defense Program Guidelines (NDPG or Defense Taiko) said that "China is widely and rapidly modernizing its military ... and is strengthening its capability for extended-range power projection. In addition, China has been expanding and intensifying its maritime activities in the surrounding waters." The Defense Taiko also called for moving away from preparing for a major war and toward a "Dynamic Defense Force" that could respond to scenarios short of a war with "readiness for an immediate and seamless response to contingencies." Confrontation between Japanese and Chinese coast guards have been occurring since 2012.[76] With their wealth of economic, military, and political might, Japan can become a problem for China in a military and diplomatic sense.

While Japan has been willing to give other countries a hand in helping develop the capabilities to stop Chinese aggression, they, like all nations, lobby for their own interests. In this case it is their need for energy security during a standoff between The United States and China. Scholars Noraini Zulkifli and Vivian Louise Forbes attribute this to Japan's status as a major user of both nuclear energy and fossil fuels. Japan is the world's third largest producer of nuclear power, after America and France; the third biggest oil consumer after America and China; the second largest importer of coal; and the largest importer of liquefied natural gas (LNG). Zulkifli and Forbes write that India and Japan have deepened their ties together, in response to their territorial

---

76 Midford, Paul (Paul Midford is an American political scientist who specializes in Japanese foreign policy. He teaches at the Meiji Gakuin University in Japan.). "Japan's Approach to Maritime Security in the South China Sea." :Japan's Approach to Maritime Security in the South China Sea on JSTOR. *Asian Survey*, Vol. 55, No. 3 (May/June 2015), pp. 525-547. Published by University of California Press. Oakland, California, U.S.A. June 2015. 525-529, 535-540, 542-543.

disputes with China. Going as far back as October 2010, Japan and India signed the "Joint Statement Vision for Japan-India Strategic and Global Partnership in the Next Decade" and a "Comprehensive Economic Partnership Agreement."

With India's naval range from the Strait of Hormuz to Singapore, this partnership benefits Japan greatly. Forbes and Zulkifli write that in September 2011, Japan and the Philippines also held talks to resolve territorial disputes in the South China Sea. In another meeting in June 2013, the Philippine government drafted plans to have Americans spend more time on military bases in the Philippines, with the same terms being offered to Japan. Japan is a strategic partner to the Philippines.[77] Japan's diplomatic ties to China's counter claimants could create a coalition if Chinese aggression reaches a fever pitch.

As hard as it is for America and China to share the world stage with each other, it has become just as hard for Japan and China to share the stage of the South Pacific. Ji Miao claims that while both Japan and China see themselves as status-quo powers, yet each accuses the other of challenging the status quo, and both are protecting their territorial claims in the East China Sea. Ji writes that China claims that Japan took the Senkakus after the first Sino-Japanese War (1894-1895). That makes the islands' history an enduring memory of Japan's involvement in China's century of humiliation. This is reinforced by Chinese Primier Li Keqiang who said in 2014 that China will not allow any country "to reverse the course of history." This is not helped by the fact that "According to a survey jointly conducted by *China Daily* and Genro NPO (in 2017), 93 percent of Japanese and 86.8 percent of

77 Zulkifli, Noraini (works at the Department of International Relations, Security & Law, National Defense University of Malaysia), Forbes, Vivian Louise (University of Western Australia · Australian Centre for Geomechanics [ACG] in Perth, Australia). "Japan and the South China Sea." :Japan and the South China Sea | Zulkifli | Andalas Journal of International Studies (AJIS) (unand.ac.id) *Andalas Journal of International Studies* Volume 5 NO.1 Mei Tahun 2016. Published by Andalas Institute of International Studies at Andalas University. Padang, Indonesia. May 2016. 63-64, 76-78.

Chinese don't have positive feeling for each other, the highest since the survey started in 2005. Only 6.8 percent of Japanese and 11.3 percent of Chinese have positive feeling for each other." Despite these sentiments, Japan and China have been willing to share the Pacific and world stage as both countries are in ASEAN+3 and G20.[78] Though China and Japan are not on good terms, they are willing to be at peace, if it's mutually beneficial.

The two nations' complicated relationship is further evidenced by their recent history with each other. Lam Peng Er mentions that, despite the tension China and Japan have with each other now, they have a history of cooperation. As part of their reparations for WW2, Japan offered aid to help develop China. From 1979 to 2016, China received 3.3 trillion yen in loan aid, 157.2 billion yen in grant aid and 181.7 billion yen in technical cooperation. Further, 23,000 Japanese companies operate in China. Tourism has also boomed between the two countries, with Japan handing out 3.78 million visas to Chinese nationals and nearly 2.5 million Japanese people visiting China in 2015. In 2014, more than 95,000 Chinese students studied in Japan, and over 15,000 Japanese studied in China. Both nations helped each other recover from natural disasters. Japan sent humanitarian aid after the 2008 Szechuan earthquake. And China sent humanitarian aid after the 2011 earthquake and tsunami hit Japan. Both Japan and China participated in the fight against piracy in East Somalia and in UN peacekeeping operations in Sudan.

However, in 2015, China's defense spending figure was $146 billion. Japan's military budget in the same year was $42 billion. Both China and Japan have a clash of nationalism. Legitimacy for China relies on

---

78 Miao, Ji (Associate Professor & Senior Research Fellow, Institute of Asian Studies, Center for China-Japan-Korea Trilateral Cooperation, China Foreign Affairs University), "Securing the Status Quo: Contesting Interests and Discourses in the East and South China Sea and Sino-Japan Relations.":159504208.pdf (core.ac.uk). *Journal of Asia-Pacific Studies* (Waseda University) No. 30 (January 2018). Published by the institute of Asia-Pacific Studies at Waseda University. Tokyo, Japan. January 2018. 17-18, 22-23.

economic growth and nationalism (which is usually anti-Japanese). Such nationalism is what sparks the dispute over the Senkakus. With the combination of national decline, and perception of China, Japanese nationalism also has been on the rise, creating a nationalist clash between China and Japan.[79] While China and Japan are currently at odds with each other, they are aware of the benefits they receive by doing business with each other.

China's upsurge in nationalism coincides with Japan strengthening its defensive military. Lionel P. Fatton brings up that ever since Abe's rise to the Prime Minister position in December 2012, the Japanese armed forces have been growing more autonomous from their American counterparts. Fatton claims this new policy was developed in the early 2010s to counter a rising China. This recent rise of autonomy comes after decades of dependence on the American military. The prospect of Japan becoming a frontline nation in a war against China is growing. The past "spear and shield" structure (where Japan would take a defensive stance, and America would use military bases in Japan to project power) is not sustainable under Japan's new security policy. Fatton contends that there is doubt in Japan about America's commitment to the South China Sea.

This fear became realized when American marines were moved from Japan to Australia, Guam, and Hawaii. This is not helped by the fact that Chinese ballistic and cruise missiles threaten American troops in East Asia. Fatton writes: "US Secretary of Defense Chuck Hagel tellingly stated at the Shangri-La defense summit in May 2014 that one of the United States' priorities was 'enhancing the capabilities of our allies and partners to provide security for themselves and the

---

79 Peng Er, Lam (political scientist, obtained his PhD from Columbia University. His publications have appeared in international journals such as the *Pacific Affairs*, *Asian Survey*, *Asian Affairs*, *Japan Forum* and *Government and Opposition: An International Journal of Comparative Politics*.). *China-Japan Relations in the 21st Century: Antagonism Despite Interdependency*. :China-Japan Relations in the 21st Century: Antagonism Despite Interdependency - Google Books. Published by Springer Nature. Singapore. 2017. 4-5, 7-9.

(Asia-Pacific) region.'" For Japan, that means that America will take a more supporting role. The main benefit of this new strategy is that Japan will become self-reliant in terms of security. However, this will be costly in both an economic and political sense. Japan's autonomy led to the creation of the "Dynamic Joint Defense Force" and the "Amphibious Rapid Deployment Brigade." Fatton goes on to write, "Japan has so far relied chiefly on US Marines for the defense of the Ryūkyū Arc and nearby remote islands, including the Senkaku/Diaoyu. The creation of the amphibious brigade ... reflects Japan's growing military autonomy but also mistrust of American security commitments in the face of an increasingly assertive China."

The first meeting of the Japanese and ASEAN defense ministers saw the lift of the 40-year-old ban on arms exports with three loose conditions on these exports: "the arms transfer shall not violate United Nations resolutions; it shall not violate Japan's international obligations; the recipient country shall not be party to a conflict in which the United Nations is involved." Since 2015, Japan has hosted training for Southeast Asian coast guards at the Coast Guard Academy in Hiroshima and the National Graduate Institute for Policy Studies in Tokyo.[80] With Japan becoming more active in the military field, it shows a shift of strategy to fight China in the scenario of war.

The reforms to the Japanese military were from reforms in legislation. This is brought up by Sheila A. Smith, who explains that Article 9 of Japan's Constitution entails the demilitarization of the nation (specifically renouncing their right to declare war on a nation). Article 9 came about as a result of Japan's actions during World War II. Debate over Article 9 has occurred over the decades, but in 2017 Abe proposed adding a paragraph to make the Self Defense Force (SDF)

---

80 Fatton, Lionel P., 'Japan is back': Autonomy and balancing amidst an unstable China–U.S.–Japan triangle": "Japan is back": Autonomy and balancing amidst an unstable China–U.S.–Japan triangle - Fatton - 2018 - Asia & the Pacific Policy Studies - Wiley Online Library, *Asia & the Pacific Policy Studies* Volume 5, Issue 2 Pages 165-386, Published by the Crawford School of Public Policy at the Australian National University in Canberra, Australia, May 2018. 264-266, 269-270, 273-274.

constitutional. This allows Japanese leaders to use Article 9 to state their intention of making a "proactive contribution to peace." Smith adds that as the possibility of fighting a war grows bigger every year, Japan must decide to either have its military integrated into a combined command with the American military, or build up their own capacity to deter or repel an aggressor without help.[81] The decades long process of legislative reform provides a perspective on how Chinese aggression has affected the nations around China.

While America and other nations in the QUAD remain Japan's biggest partners, Japan also looks toward the neutral ASEAN for another part of its defense and deterrence against China. The relationship between Japan and ASEAN is explained by Takashi Shiraishi, the chancellor of the Prefectural University of Kumamoto, and Takaaki Kojima, a former Japanese Ambassador to Singapore in their book *ASEAN-Japan Relations*. Shiraishi and Kojima point out that the strengthening of relations with ASEAN began with Abe in 2012. The following year, Abe visited Vietnam, Thailand, Indonesia, Malaysia, Singapore, and the Philippines. Japan was once the largest donor of development assistance to ASEAN, providing aid for economic and social infrastructure such as roads, ports, power stations, water and sewage systems, and human resource development. Japan also helped ASEAN nations when they were hit with the 1997-98 economic crisis. In December 2003, the 30th anniversary of the relationship between Japan and ASEAN took place. The leaders and foreign ministers of Japan and ASEAN met in Tokyo for a "commemorative summit." According to a summary made by the Japanese government: "Prime Minister Junichiro Koizumi clearly stated that Japan would continue to regard its relationship with ASEAN as very important and confirmed that Japan and ASEAN should further develop their relations as sincere

81 Smith, Sheila A. (John E. Merow senior fellow for Asia-Pacific studies at the Council on Foreign Relations [CFR]). *Japan Rearmed: The Politics of Military Power.* :Japan Rearmed: The Politics of Military Power - Sheila A. Smith - Google Books. Published by Harvard University Press. Cambridge, Massachusetts. 2019. 2, 5, 9, 12, 15.

and open partners that 'act together (and) advance together.'" Then Indonesian President Megawati Sukarnoputri spoke for ASEAN by saying, "In the past three decades, ASEAN and Japan have developed a mutually beneficial partnership and cooperation which continue to deepen and widen. Let me reaffirm that the ASEAN-Japan long-standing relations and cooperation have contributed significantly to the peace and prosperity of the people in Southeast Asia as well as in the wider region of East Asia and the Asia-Pacific."[82] The relationship between Japan and ASEAN could draw the organization further into a partnership with the QUAD.

While Shinzo Abe is no longer in power in Japan, his impact on the country remains at center stage, especially when it comes to countering Chinese aggression. Gilbert Rozman writes that Abe, as Japan's longest-serving prime minister after World War II, was able to focus on his country's international presence. During his first visit to Washington, D.C. in 2013, Abe said he wanted Japan to maintain its status as a "first-rate country." He told his audience that Japan will be "the guardian of the commons," or a nation that will uphold international norms and rules. Rozman writes that "Abe's message to China has remained consistent ... he took every occasion he could to reinforce his message that Japan would not compromise its position on the sovereignty of the Senkaku Islands and that it would counter Chinese attempts to change the status quo by coercive measures ... while he was consistent in his message that Japan would not tolerate Chinese aggressive behavior to assert its sovereignty claims, Abe also made it clear that Japan was willing to engage with China in the areas that are mutually beneficial."

Abe had trouble with Russia and South Korea, however. He could not get the Russian government to budge on their claims on

---

82 Shiraishi, Takashi (Chancellor, Prefectural University of Kumamoto; Professor Emeritus, National Graduate Institute for Policy Studies [GRIPS], Japan), Kojima, Takaaki (Visiting Senior Fellow at ISEAS and the former Japanese Ambassador to Singapore). *ASEAN-Japan Relations*. :ASEAN-Japan Relations - Google Books. Published by ISEAS. Singapore. 2014. II, 2-4, 17.

the Kuril Islands (Islands north of Japan). With South Korea, Abe had to overcome bitter memories of atrocities perpetrated in World War II. Rozman writes, "(The) Japanese do not deny that atrocities took place, but they are offended when they are laid at the feet of current generations." It does not help matters that both Japan and South Korea signed the 1965 Treaty on Basic Relations, which says in Article II "(The) problems … and interests … and the claims between the High Contracting Parties (Japan and South Korea) and between their peoples … have been settled completely and finally." In 2014, Abe called South Korea "our most important neighboring country, with which we share fundamental values and common interests. Good relations between Japan and the Republic of Korea (ROK) are indispensable not only for our two countries but also for the peace and prosperity of East Asia."

Rozman claims that China could use this division between Japan and South Korea to their advantage, and points out that Japan has conflicting opinions on China's economic rise. Japan has benefitted from the economic opportunity with China and the two nations often seem to be economic and environmental partners. It has been the Sino-centrism of the Xi Administration that has made Japan wary. In 2018, Japanese companies invested nearly $11 billion, or a 50 percent increase from 2010. Japan is tapping into the $42 trillion financial industry in China. As Rozman says, "By 2018, China was Japan's largest goods-trading partner, with $317 billion in two-way flows, representing over 21 percent of total Japanese merchandise trade. … At the end of 2018, China was the fourth-largest destination for Japanese foreign direct investment (FDI), with $124 billion in total FDI stock. As of May 2019, 42 years after Beijing granted the first business permit to a Japanese company, there were 13,685 Japanese companies doing business in China, up from 10,778 in 2010, according to survey data." And, "As many as 119 Chinese businesses made Fortune's 2019 Global 500 index of the largest companies, up from only 5 in the 1999 edition. Over that time, Japan's share fell by more than half, from 107 in 1999 to

52 in 2019."[83] What Abe did as minister is remilitarize to keep up with an ever growing China.

While the scenario of the South China Sea War would primarily be a fight for Taiwan, Japan and other nations fighting China would have to defend territory that China claims is theirs. For Japan, it is the Senkaku Islands. The Senkaku issue is discussed in detail by Edmund J. Burke, Timothy R. Heath, Jeffery W. Hornung, Logan Ma, Lyle J. Morris, and Michael S. Chase. The authors explain the rivalry between Japan and China, which has flared up recently with air patrols going on between the two nations over the islands. In 2016 the JASDF scrambled 1,168 aircraft, or 73 percent of the JASDF, over the presence of Chinese aircraft over the Senkakus. Japan claims that not only do the Chinese want the Senkakus because of their strategic position in the first island chain, but also because it was discovered in the 1970s that there are potential energy deposits on the islands. Though China could overwhelm the JASDF in a war, the authors write that another potential strategy would be to use economic coercion to take the Senkakus without firing a shot. This could take the form of reducing the number of Chinese tourists going to Japan or reducing rare earth mineral exports.[84] The first island chain will be the first line of defense, where most of the fighting would take place in a scenario of war.

---

83 Rozman, Gilbert (an American sociologist specializing in Asian studies). Joint U.S.-Korea Academic Studies: East Asian Leaders' Geopolitical Frameworks, New National Identity Impact, And Rising Economic Concerns With China 2020|Vol.31. :kei_jointus-korea_2020_final_digital_0.pdf (keia.org). Published by Korea Economic Institute of America, Washington, D.C., U.S.A. 2020. 66, 72, 76, 188, 191, 194, 210, 237, 243.

84 Burke, Edmund J. (senior international defense researcher at RAND. B.A. in government, Georgetown University), Heath, Timothy R. (Timothy R. Heath is a senior international defense researcher at the RAND Corporation. Prior to joining RAND, Heath had over fifteen years of experience in the U.S. government researching and analyzing military and political topics related to China.), Hornung, Jeffery W. (senior political scientist at the RAND Corporation. He specializes in Japanese security and foreign policies, East Asian security issues, and U.S. foreign and defense policies in the Indo-Pacific region, including its alliances. Prior to joining RAND in April 2017, Hornung was the fellow for the Security and Foreign Affairs Program

As Japan rearms itself post-World War II, it seeks to protect its interests just as any other nation would. Their wariness of the Chinese and the Russians has caused the Japanese government to maintain relations with America. This decision was made as America and Japan have enjoyed a military relationship going back to the late 1940s. However, Japan's desire to invest in Chinese markets and their generational tension with South Korea has left the Japanese government with the ability to take small steps to how they position themselves between America and China. But given the aggressive claims of China, Japan has a keen interest to keep Chinese ambitions within the borders of China.

---

at Sasakawa USA from 2015 until 2017.) , Ma, Logan (adjunct research assistant at RAND. He is currently a master's student at the Johns Hopkins School of Advanced International Studies.), Morris, Lyle J. (Senior Fellow for Foreign Policy and National Security at Asia Society Policy Institute's Center for China Analysis. Prior to joining ASPI, Lyle was a senior policy analyst at the RAND Corporation leading projects on Chinese military modernization and Asia-Pacific security from 2011-2022.), Chase, Michael S. (became the Deputy Assistant Secretary of Defense for China on 8 February 2021 and assumed the responsibilities for Taiwan and Mongolia in October 2022. He was previously a senior political scientist at RAND where his research focused on Chinese military modernization, China's approach to strategic deterrence, Taiwan's defense policy, and Indo-Pacific security issues). *China's Military Activities in the East China Sea: Implications for Japan's Air Self-Defense Force.* :China's Military Activities in the East China Sea: Implications for Japan's Air Self-Defense Force (rand.org).Published by the RAND Corporation. Santa Monica, California, U.S.A. 2018. VI, 1, 6-8.

Chapter 6: South Korea: Stuck Between the Dragon's Den and the Eagle's Nest

A year after the Chinese Civil War, Communist North Korea and Democratic South Korea clashed in a 3-year war, which brought China and America to blows as they supported North Korea and South Korea, respectively. The Korean War (1950-1953) concluded with a ceasefire between the two Koreas, which has led to a 70-year standoff, with China and America having vested interest in the fate of the Korean Peninsula. South Korea has dealt with the constant threat of North Korean bombing and China's economic power in the country. Not only that, but America has military bases in South Korea and is willing to defend their ally in the peninsula. This has left the South Korean government to perform a balancing act in their relations with America and China.

While they have remained neutral over the decades, the geographic location of South Korea makes it difficult for them to refrain from taking sides in the South China Sea issue. Terrence Roehrig points out that despite not being located directly on the South China Sea, South Korea is being pressured to take a side for or against China's actions in the region. This, as Roehrig states, is because South Korea is the seventh largest trading state in the world. In 2015 alone, South Korean trade with Vietnam, Singapore, Indonesia, Malaysia, and the Philippines reached to the tens of billions of US dollars each. A war or a blockade of the South China Sea would impact South Korea negatively. Much like other South China Sea nations, South Korea has territorial claims that put them in dispute with neighboring countries. More specifically, South Korea's claims put them at odds with China, Japan, and North Korea. With China, South Korea has overlapping claims over their EEZs in the Yellow or West Sea.

South Korea also has engaged in disputes involving illegal fishing from Chinese fishing boats. This causes problems such as overfishing and pollution in South Korea's EEZ. Violence has been common between Chinese fishing boats and the South Korean Coast Guard. In 2015, South Korea seized more than 600 Chinese fishing boats. The final issue is that South Korea has feuded with China over a reef called by the South Koreans "Ieodo" and by the Chinese "Suyan Rock." In 2003, South Korea built the Ieodo Ocean Research Station on the reef to collect data on ocean currents, fishing, weather, and climate change. South Korea has argued that it was permissible to build the research facility under UNCLOS Article 60 (the right of a coastal nation to build infrastructure such as artificial islands and regulate drilling in their EEZ) and UNCLOS Article 80 (the right of a state to enforce customs, tax, immigration and pollution infringement laws in the land or coastal territory of said state).

China was displeased by this, and 10 years later China established an air defense identification zone (ADIZ), which covered Ieodo. In November of that year, South Korea expanded its own ADIZ which included the reef, albeit after notifying Japan, China, and America. As one analyst put it, "Until now (2015), Seoul has avoided taking sides in territorial disputes involving China and other countries in the region. Now that it faces its own territorial tug of war with China [over Ieodo], South Korea may find it difficult to remain neutral in such disputes in order to counteract China's growing assertiveness in claiming territories in the region."

The dispute between South Korea and Japan is over a cluster of islets called the "Dokdo" to South Korea and "Takeshima" to Japan. The islets are largely known for their fishing and the possibility of gas and oil deposits. Ever since the late 1800s, the islets have remained a flash point for Japanese-Korean relations. However, at present this leads to legal battles rather than military battles. The maritime tension between North and South Korea is over 5 islands called the Northwestern Islands (NWI) and the Northern Limit Line (NLL), the maritime line between the Koreas.

Roehrig claims that since both China and America are important to South Korea's future, the South Korean government will have to navigate between the two great powers. China is South Korea's top trader, beating the combined trade from America and Japan. Because it represents 25 percent of South Korea's trade, China can—and has—used its economic leverage over South Korea. When South Korea deployed the American Terminal High Altitude Area Defense (THAAD) missile defense battery, China retaliated by imposing restrictions on travel and tourism to South Korea and cancelling all K-Pop concerts. South Korea also sees China as an important partner in addressing the problem of nuclear weapons in North Korea because China is North Korea's economic and political lifeline. After becoming president of South Korea in 2013, Park Geun-hye met with Xi Jinping on several occasions to discuss many topics, including North Korea. On one of these trips, both Park and Xi warned North Korea to refrain on taking any action that would cause a regional conflict. Roehrig writes that "A Chinese press report noted that both leaders expressed a willingness to cooperate in achieving denuclearization on the Korean Peninsula along with communicating and coordinating on regional and international affairs."

Despite this, North Korea conducted its fourth nuclear weapons test in January 2016. And the lack of a response from China was less than what Park hoped for. While the THAAD was delayed, after China's lukewarm response, THAAD was back on the table for South Korea. This enraged the Chinese. Their ambassador to South Korea, Qiu Guohong, warned that relations between the two nations could be "'destroyed in an instant' if the THAAD system were deployed to Korea." Roehrig writes that "a Park Administration spokesman fired back: 'This is a matter we will decide upon according to our own security and national interests. The Chinese had better recognize this point.' A senior official at the ROK Foreign Ministry who wanted anonymity chimed in further that China should 'look into the root of the problem (North Korea's nuclear weapon and ballistic missile programs) if it really wants to raise an issue with it.'" On July 8, 2016, the South Korean Defense Ministry announced an agreement with America to deploy THAAD. Chinese Foreign Minister Wang Yi said, "The recent move by the South Korean side has harmed the foundation of mutual trust between the two countries. THAAD is most certainly not a simple technical issue, but an out-and-out strategic one."

Despite debate about THAAD in her own country, Park and her administration were committed to keeping that system because, in her words, "Growing nuclear and missile threats are a very critical issue where the future of the Republic of Korea and the lives of our people are at stake. As president, I have the obligation to protect our people and nation." Park added that "THAAD will not target any country other than North Korea and will not encroach upon the security interests of any third country. (We) have no reason to do so." China was also mute when North Korea conducted their fifth test in September 2016, which included ballistic missiles. During the 2016 G20 summit in China, Xi reiterated that "China and South Korea should make efforts to get their bilateral ties back on track toward stable development" and that "close neighbors with common interests ... should hold dear the foundation of their political cooperation and get over difficulties and challenges."

South Korean Finance Minister Yoo Il-ho argued, "There will be no large-scale retaliation from China. We speculate that (the Chinese government) will separate economics with politics. Although we are concerned, it is unlikely that economic relations between South Korea and China will see a sudden decline. We will try to convince China that this is a political issue and not an economic one." Despite their grievances, South Korea still sees China as the only nation that can hold back North Korea. However, Roehrig notes that polling from the Asian Institute shows that from 2010-2014, 85 percent of South Koreans believed that the Alliance with America was necessary for the security of South Korea. Roehrig writes that "In October 2013, the alliance announced a new 'Tailored Deterrence Strategy' (TDS) to counteract North Korea's nuclear, chemical, and biological weapon capabilities. Though the details are classified, the TDS contains plans for further integration of US and ROK forces for a joint response to the Democratic People's Republic of Korea (DPRK) threats, including the possibility of preemptive strikes should a North Korean attack appear imminent." An exercise in 2016 showed the alliance uses a different strategy called "4D" (detect, disrupt, destroy, and defend): a pre-emptive strike plan in which the alliance would destroy and secure DPRK chemical and nuclear weapon assets during a conflict to prevent another state or terrorist organization from accessing those weapons.

Roehrig recounts the following comments made by then-President Barack Obama during a press summit: "So, there's no contradiction between the Republic of Korea having good relations with us, being a central part of our alliance, and having a good relationship—good relations with China. I think as I communicated to President Park, the only thing that we're going to continue to insist on is that we want China to abide by international norms and rules. And where they fail to do so, we expect the Republic of Korea to speak out on that, just as we do, because we think that both of our countries have benefitted from the international norms and rules that have been in place since the end of World War 2." Defense Minister Han Min-koo said, "The stance of the Republic of Korea is that a peaceful resolution of the South China Sea dispute and the freedom of navigation and flight should be guaranteed. The dispute should be settled peacefully through an agreement among related parties and in accordance with international standards." In 2015, Park added to Min-koo's comment with: "Korea has consistently stressed that the dispute must be peacefully resolved according to international agreements and code of conduct. China must guarantee the right of free navigation and flight."[85] Though both China and America remain pivotal to Korea's economy and military respectively, the actions of both nations impact South Korea too much to ignore.

However, given that South Korea is near China, Russia, Japan, and America, there is no clear choice as to which side South Korea will choose to back in the scenario of a South China Sea War. This complex matter that South Korea has found itself in is explained by Corey J. Hubbard and Paula Brito. Hubbard and Brito assert that when it comes to North Korea, America is the only player that has taken proactive measures in the matter. North Korea has the fourth largest ground force in the world, and South Korea has the sixth most powerful military in the world. South Korea has benefitted from American military aid since the Korean War, with 28,500 American soldiers being stationed in South Korea. Hubbard and Brito write that "Russia denounced both the DPRK's nuclear test and the THAAD system's expansion, fearing they would lead to regional hostilities." China, meanwhile, supports North Korea as it prevents America from getting a hold on the entire Korean Peninsula.

While Japan is trying to have a tri-lateral relationship with America and South Korea, there is tension over the Japanese occupation of South Korea from 1910 to 1945. In one agreement, the Japanese government agreed to pay 1 billion yen to 46 surviving Korean victims of Japan's "Comfort Women" program. After the COVID-19 pandemic, the Biden Administration increased the number of US soldiers stationed in South Korea and Japan. The authors claim that though both countries have

---

85 Roehrig, Terrence (Ph.D. Professor National Security Affairs Biography Professor of national security affairs). "Caught in the Middle: South Korea and the South China Sea Arbitration Decision." :Caught in the Middle: South Korea and the South China Sea Arbitration Decision in: Asian Yearbook of International Law, Volume 21 (2015) (brill.com), *Asian Yearbook of International Law*, Volume 21 (2015) 96-120. Published in Foundation for the Development of International Law in Asia (DILA). The Hague, Netherlands. September 17, 2017. 97-113.

grown wary of China because of COVID-19, young South Koreans see this increase as detrimental to reunification efforts with North Korea.

While America is helping South Korea in a military sense, China and Russia have been a boon to South Korea economically. China opened diplomatic ties to South Korea in 1992 and became their top trading partner in 2004. The two nations signed the China-Republic of Korea Free Trade Agreement in 2014. Since then, China has used economic pressure to discourage South Korea from allowing further American influence. Hubbard and Brito explain that since the fall of the Soviet Union, trade between Russia and South Korea has skyrocketed, and Russian investment in South Korea has increased 10-fold. Both Russia and South Korea share a common distrust of Japan due to their past. Talks between then-President Moon Jae-in and President Putin developed a cooperative energy-security relationship. That resulted in Russia's 2030 Energy Strategy, which transports liquefied gas via pipes going through North Korea. Like China, Russia is wary of America's influence on the Korean Peninsula. However, if Russia were to increase their influence on the peninsula, China would have to lessen their influence as well. This provides South Korea with what is known as the "Russian alternative," a way that South Korea could potentially curb North Korean aggression. The authors add that though Russia would help contain North Korea, that does not mean they will support denuclearization.[86] Hubbard and Brito conclude that where the US establishes itself in South Korea with its military, China and Russia use their economies to prop up their influence in the country.

A problem that South Korea shares with other East Asian nations is a declining population. Hubbard and Brito write that the combination of an aging population and low birth rates can lead to

---

86 Hubbard, Corey J. (a researcher and author affiliated with the University of Navarra), Brito, Paula (Bachelor Degree in International Relations at the University of Navarra). :ROK_SAR_2021-April_CHubbard-PMora.pdf (unav.edu). *South Korea Strategic Analysis Report/April 2021: How the Inter-Korean Conflict is an indication of the New Cold War.* Published by the Center for Global Affairs & Strategic Studies at the University of Navarra. Pamplona, Spain. 2021. 19-21, 26, 28-31, 33-34, 38.

economic stagnation in the future. The authors go as far to claim that the country's population will peak by 2030 with 52 million people, and then decline to 44 million in 2060. As a result, South Korea would lose 17 percent of its labor market. This fact has led to calls for South Korea to reform its immigration policy.[87] The decline in population could lead to a cooling of tension in the future, as the nations of East Asia could lack the manpower to fight a war.

The beneficial relationships that South Korea maintains with both America and China have caused debate within the country as to whether one partner should be favored over the other. Gilbert Rozman stated: "The question of 'China or the United States?' dominated South Korean strategic thinking from the beginning of the Lee Myung-bak administration in 2008." Rozman recounts that "A joint report by the Asan Institute for Policy Studies, Chicago Council on Global Affairs, and Korea Economic Institute at the end of 2012 showed that just over half of South Koreans (53.5 percent) viewed China as a partner rather than a rival (46.5 percent), with an overwhelming 94 percent expressing support for the US-ROK alliance." Rozman then adds that, according to the Asian Institute, South Korean favor towards China diminished from 73 percent in 2006 to 53 percent in 2011. He identifies two perspectives on post-Cold War China in South Korea. One highlights China's assertiveness, using military, economic, and political coercion to dominate the region. And the other showcases China's rise onto the global stage and importance as an economic partner for the country. China adds to the debate as they denounce America-South Korean cooperation as "Cold War thinking."

In August 2011, an editorial by Gilbert Rozman cited the president of South Korea's Society of Ieodo Research decrying Chinese behavior in the EEZ clashes with Vietnam as "imperialistic." It went further to say that "Asia's mistrust of China and fear of Beijing is based on its territorial ambition ... China's ambition should be counterbalanced by the United States as a Pacific partner to Asia-Pacific nations." In

87 Hubbard and Brito. 8, 12.

response to a Chinese aircraft-carrier trial that same month, a Korea National Defense University professor said that "Korea can secure military deterrence by reinforcing joint deterrence capacity with the United States." Rozman concludes that China's expansionism and relations with North Korea feed some suspicions for South Korea as to intruding into the Korean Peninsula.[88] Though South Korea is willing to maintain an economic relationship with China, the combination of China's aggression and America's military assistance has pushed them toward the US, rather than China.

Much like Japan, South Korea is also developing a powerful military to counter aggressive nations. Resi Aini and Febry Triantama explain the modernization of the South Korean military. Aini and Triantama note that the main point of military modernization for South Korea is to defend against a North Korea that wants to unify the Korean Peninsula with force. They write: "South Korea launched the 2020 Defense Reform program which was initiated under the President Roh Moo-hyun's administration (2003-2008) in order to increase the reliability of national defense … Both the 2020 Military reform launched by President Roh Moo-hyun and the Military Reform 2.0 under President Moon Jae-in, explicitly mentioned the government's priorities for the development of military technology and exports of the defense industry."

Another part of South Korea's military modernization is the Revolution in Military Affairs (RMA), which is an effort to make the South Korean military independent from other nations. According to the authors, by 2019, South Korea had 2,400 tanks, including their own K-1 and K-2 tanks, 2,300 self-propelled artillery such as K-9 and K-55, and 3,200 armored vehicles of various types. Aini and Triantama claim that this makes the South Korean military one of the most armored militaries in the world. They connect the modernization of South Korean military with the need for technology for national security, which cultivated techno-nationalism in South Korea, meaning a

88 Rozman. 98-100, 105-107.

country that wants to control their technology independently. They bring up while in 2008, South Korea's defense industry exports were at 1.03 million American dollars and in 2017, it reached to 3.12 million American dollars, the nation's arm sales went from 197 million Trend Indicator Value (TIV) (a measurement of military capablity) in 2010, and in 2019 that number reached to 1.06 billion TIV. To the authors, this is impressive, considering that China's arms sales TIV was at 1.15 billion at its smallest and 2.37 billion TIV at its biggest between 2010-2018, and Russia's arms sales TIV was at 5.5 billion TIV at its lowest, and 8.7 billion TIV at its highest. The authors contend that South Korea must modernize its military, because it should not depend on the US for defense.[89] The modernization of the South Korean military can help defend the country, but the country must consider additional steps before having to take on North Korea and China on the battlefield.

The relationships among South Korea, the US and China have a complex history. The politics of South Korea in the early-to-mid 2010s are explored by Chung Jae Ho, who elaborates as to China's view on South Korean presidents of the early 2010s. China was happy about the ascension of Park Geun-hye (2013-2017) to the South Korean presidency in 2013. The previous administration of Lee Myung-bak (2008-2013) prioritized South Korea's alliance with America and relegated China into a secondary or tertiary role, while North Korean provocations in 2010 put Sino-South Korean relations in an awkward

---

89 Aini, Resi (Lecturer and Researcher at the University of Indonesia), Triantama, Febry (Researcher and Lecturer with a demonstrated history of working both in the Public Policy Industry [Think Tanks] and University. Academia with a bachelor's degree in international relations from Paramadina University and a Master's Degree in International Relations focusing on International Security from the University of Indonesia with The Indonesia Endowment Funds for Education [LPDP] Scholarships.). "The Implementation of South Korea's Military Technology Reform in The Perspective of Techno-nationalism." :The Implementation of South Korea's Military Technology Reform in The Perspective of Techno-nationalism | Sospol : Jurnal Sosial Politik (umm.ac.id), *Sospol Jurnal Sosial Politik* Vol. 7 No. 1 (2021): January-June. Published by the University of Muhammadiyah Malang. Malang, Indonesia. June 29, 2021. 64-66, 68, 71-73.

state for two years. Geun-hye's trips to China from 2013 to 2014 were calculated to show a greater relationship between the two countries. Xi's visit to South Korea in 2014 produced a free trade agreement (FTA), and led to negotiations about Ieodo

The years 2013-2015 marked a great era for Sino-South Korea's relationship. Jae Ho writes: "Seoul's hesitation to join the negotiations for the US-centered Trans-Pacific Partnership (TPP) agreement was one indicator although Seoul had its own rationale—having signed FTAs with 10 of the 12 countries in the framework, joining the TPP would not bring much marginal utility. Second, despite Washington's opposition, Seoul's decision to join the China-initiated Asian Infrastructure Investment Bank (AIIB) in 2014 was deemed as yet another sign of South Korea tilting more toward China." This relationship, however, as Jae Ho discusses, was shattered in 2016, due to the THAAD controversy, in which South Korea was trying to establish an American-made surface-to-air missile defense system. THAAD would give South Korea the ability to defend itself from North Korean missile attacks. China felt like they were misled from 2014 to early 2016, when the South Korean government followed the "Three-Nos" policy (no request from the US; no consultation with Washington; and no decision whatsoever about THAAD). "On February 12, 2016, Chinese foreign minister Wang Yi, in an interview with Reuters, cited an old Chinese saying to the effect that 'South Korea is dancing with a sword to help the United States, but her real intention is to kill China,'" according to Jae Ho. The Chinese rejected the explanation that THAAD was aimed at North Korea and not China. When North Korea launched a missile on July 4th, 2017, South Korea decided to not withdraw the THAAD system. That was supported by popular opinion since 57 percent of South Koreans want to keep the THAAD system, whereas 27 percent are opposed.[90] Although China and South

---

90 Ho, Chung Jae (a professor of international relations and director of the Institute for Chinese Studies at Seoul National University, was nominated by South Korean President Yoon Suk-yeol as the new ambassador to China in 2022). "South

Korea remain partners, North Korea remains a controversial issue that neither side will back down from soon.

The United States and South Korea do have good relations, but the relationship between South Korea and Japan remains rocky at best. Prakash Panneerselvam brings up that in 2018, the South Korean Supreme Court took the case of compensation of South Korean workers forced to work in Japanese companies during World War II. The verdict of the case ordered Mitsubishi to pay compensation for their use of forced labor of South Koreans. Nippon Steel & Sumitomo Metal Corp (NSSM) were also ordered to pay compensation for forced labor. This brought up wartime memories in the two nations, and the following year a trade war broke out between South Korean and Japan. Nearly 80 Japanese companies are facing similar charges by the South Korean Court. Panneerselvam writes that the Japanese government believes this and similar cases undermine the legal relationship between themselves and the South Koreans. Japanese leaders believe "that every bilateral problem pertaining to the colonial past has already been settled by virtue of the treaty (of Basic Relations in 1965). Article II of the Agreement (treaty of 1965) and (the continuing litigation) inflicts unjustifiable damage and costs on the said Japanese company." Japan's Shinzo Abe said to reporters that "the ruling is impossible in terms of international law." Lee Nak-yon, then prime minister of South Korea, said the Japanese remarks were "aggressive," and were "neither legitimate nor wise."

Panneerselvam claims that experts fear that this rhetoric could make Japanese companies leave South Korea. In response to the verdict and the South Korean government's stance, Japan imposed sanctions on three essential chemicals used by such tech giants as Samsung and LG. These three chemicals are fluorinated polyimides, which are used

---

Korea's Strategic Approach to China (or Lack of It)." Joint US-Korea Economic Studies. :jukas_1.4_south_koreas_strategic_approach_to_china.pdf (keia.org). Published by the Korean Economic Institute. Washington, D.C., U.S.A. August 13, 2018. 74-79, 81.

in smartphone displays; photoresists, which are used to transfer circuit patterns onto semiconductor wafers; and hydrogen fluoride, which is used as an etching gas in the chip-making process. Japan produces 80 percent of the world's fluorinated polyimides and photoresists. On August 22, 2019, South Korea announced that it would not renew its involvement in the bilateral General Security of Military Information Agreement (GSOMIA). This was a blow to the America-Japan-South Korea trilateral security cooperation against North Korea. However, as North Korea remains a persistent issue, South Korea reversed its decision in November that same year.[91] South Korea is willing to work with the Americans and their partners, including Japan, on security issues. The combination of a vital relationship with America and an aggressive North Korea are the glue that holds these three countries together.

South Korea's internal views on China, America, and Japan are split between two groups: the progressives and the conservatives. The aforementioned split is explained by Gilbert Rozman, who writes that "Progressives in South Korea generally preferred the idea of making positive steps toward inter-Korean collaboration with little regard to North Korean denuclearization. They expressed hope for the possibility that North Korea would denuclearize itself if advanced measures of inter-Korean cooperation were adopted. In contrast, conservatives remained highly suspicious of Kim Jong Un's commitment to North Korean denuclearization, arguing that Kim would never drop his nuclear weapons voluntarily. Even though Kim may send positive signals toward denuclearizing North Korea, conservative groups think that such talk can be highly deceptive. If that is the case, how to defend

---

91 Panneerselvam, Prakash (an Assistant Professor with the International Strategic & Security Studies Programme, National Institute of Advanced Studies [NIAS], Bengaluru, India). "Changing Dynamics of Regional Trade and Cooperation: A Case of Japan and South Korea." :Changing Dynamics of Regional Trade and Cooperation: A Case of Japan and South Korea | *Artha Journal of Social Sciences* (christuniversity.in). VOL. 19 NO. 4 (2020): *ARTHA Journal Of Social Sciences*. Published by Christ University, Bangalore, India. 2020. 1-2, 4-6, 8-12.

South Korean security from an increasing North Korean military threat assumes priority."

Rozman claims that "Instead of naively advancing inter-Korean collaboration, conservatives argue that tightening security cooperation with the United States should remain the priority. Along with security cooperation with the United States, cooperation with Japan, a US ally, sounds reasonable from the conservative standpoint." Rozman writes that Progressives have been generous to China, but they have been critical of Japan. Their anger towards Japan stems from security issues, diplomatic struggles, and human rights concerns. He notes that South Korean people in their 20s and 30s have more favorable views to Japan, while the older generation continues to have negative views. Rozman then writes "pro-Japanese collaborators, whose descendants are mostly found in conservative circles, are depicted as illegitimate people who occupied higher social positions despite the will of the Korean masses."[92] The political split in South Korea could prove to be the deciding factor as to what side the country will take in the scenario of a war.

South Korea has taken notice of Taiwan and how a war between China and Taiwan would affect them. Seong-Hyon Lee explains that South Korean observers see Taiwan as one of four hotspots which can cause a war between China and America. The other three being the Senkaku Islands, the Korean Peninsula, and the South China Sea. 90 percent of South Korea's maritime trade goes through the Taiwan Strait and South China Sea. Their security and economic ties with America and China respectively have led South Korea to maintain a cautious diplomatic stance. Lee mentions that during Chinese aggression, South Korean President Yoon Suk Yeol endorsed the Camp David joint declaration (a document released by the White House which accused China of "dangerous and aggressive behavior" in the South China Sea.

He reports on a November 1, 2023, workshop held by the Council on Foreign Relations and titled "US-South Korea Policy Coordination

92 Rozman. 173-174, 176, 178.

Toward China on Maritime Security." Lee described his takeaways from the workshop at great length: "Despite the stated urgency of the matter, and even as the Taiwan issue has become a significant policy priority for the Joe Biden administration, it has not been fully integrated into South Korea's normative security considerations, indicating Seoul's cautious stance amid volatile regional dynamics. South Korea's approach to tensions across the Taiwan Strait is contingent on several pivotal factors. At its core lies Seoul's confidence in its deterrent capabilities against Pyongyang, the capital of North Korea, particularly given North Korea's de facto status as a nuclear state … China's status as South Korea's key trading partner—combined with its geographical and geopolitical proximity, including China's leverage over North Korea—casts a long shadow over Seoul's potential role in any Taiwan contingency.

"Since the Korean War, South Korea has not faced direct military conflict with China, positioning any involvement in a Taiwan crisis as an unprecedented venture fraught with considerable risks to national security. … South Korea's potential involvement in a Taiwan contingency would likely hinge on the degree of the United States' own attention toward the situation and the strength of the Washington-Seoul alliance. This situation mirrors the dynamics that were at play when Russia invaded Ukraine, during which the level of engagement by US allies was primarily steered by Washington's leadership and its orchestration of allied actions, rather than direct bilateral interactions with Ukraine … Washington could facilitate improvements by subtly encouraging interaction, leveraging its convening power, and inviting Seoul and Taipei to the same gatherings.

"It is a common misconception that countries in the same region, both reliant on the United States for security, will naturally bond and strengthen ties. The relationship between South Korea and Japan proves otherwise, a point that has often frustrated US policymakers. Diplomatic relations do not spontaneously improve; they require effort and careful navigation. For instance, the semiconductor industry is an area in which South Korea and Taiwan are fierce competitors. As South Korea positions itself as a 'global pivotal power,' it needs to embody this role. It can no longer justify its limited engagement with Taiwan by citing preoccupations with North Korea, fears of Chinese retaliation, or domestic political schedule, such as the upcoming legislative elections in April this year."

Lee suggests that America and South Korea could collaborate on maritime surveillance and defense technologies. He adds that the two nations also need to make a joint strategic framework that addresses concerns from the Taiwan Strait and the South China Sea. America needs to further trilateral relations with Japan and South Korea. America and South Korea should enhance relations with ASEAN to build up a resilience to coercive actions in the South China Sea. They can reaffirm international maritime laws, especially UNCLOS, and promote stronger economic ties with other Southeast Asian nations to mitigate the over-reliance on China.[93] South Korea could help with deterring Chinese aggression against Taiwan, but this would leave them open to both China and North Korea, thus explaining South Korea's neutrality on the Taiwan issue.

However, the border crisis between the two Koreas might move South Korea to a pro-American approach. Hyonshee Shin writes that on April 16, 2023, "South Korea fired warning shots toward a North Korean vessel that crossed the maritime border, South Korea's military said on Sunday, a day after the incident that further raised tension over the North's missile tests." Shin cites South Korea's joint chiefs of staff (JCS), who said the country's military "fired warning shots and broadcast warnings to expel a North Korean patrol boat that breached the Northern Limit Line (NLL), the de facto sea boundary, around 11 a.m. (0200 GMT) on Saturday." And "'Our military maintains decisive

93 Lee, Seong-Hyon (senior fellow at the George H. W. Bush Foundation for U.S.-China Relations and visiting scholar at the Harvard University Asia Center). "U.S.-South Korea Coordination Toward China: Maritime Security." Council on Foreign Relations. January 4, 2024. https://www.cfr.org/blog/us-south-korea-coordination-toward-china-maritime-security.

battle posture while monitoring the enemy's movements in preparation for potential provocations regarding NLL violations by North Korean patrol boats,' the JCS said in a statement on Sunday." Shin explained that this incident came at the aftermath of North Korea doing a missile test the previous day. Since the 1990s, North Korea has disputed the NLL and argued that it should be pushed farther north.

In October of 2022, the two Koreas fired shots in their western waters accusing each other of breaching the sea borders. In the same week of the missile test and subsequent sea skirmish, Kim Jong Un ordered a 'more practical and offensive' manner to counter what the isolated country called moves of aggression by the United States and South Korea." Shin explains that "Pyongyang threatened military action as South Korea and US forces have been carrying out their annual springtime exercises since March, calling them a rehearsal for nuclear war. South Korea's air force said on Sunday it would stage combined drills with the US Air Force and Marines from Monday until April 28. The exercises would mobilize 110 aircraft, including South Korea's F-35 and F-15 fighters and American F-16 jets and KC-135 aerial refueling tankers, as well as 1,400 troops from both sides, it said. ... Seoul and Washington say their drills are defensive and aimed at deterring the North."[94] Though South Korea wants to remain neutral, their hostile maritime border with North Korea could move the nation to side more closely with America than with China.

Though China and South Korea enjoy close relations, China's aggression is driving a wedge between the two countries. Gilbert Rozman explains this growing divide between two of the most powerful nations in East Asia. Since 2010, there has been a debate within South Korea about Chinese assertiveness. Both China and South Korea see North Korea as a "little brother." The conflicting realities of South Korea being a military ally to America and an economic partner of

94 Shin, Hyonshee (Journalist for Reuters). " South Korea fires warning shots after North's boat crosses sea border." Reuters. https://www.reuters.com/world/asia-pacific/south-korea-fires-warning-shots-after-norths-boat-crosses-sea-border-2023-04-16/.

China drives this debate as the two major powers grow warier of each other. Rozman writes: "A joint report by the Asan Institute for Policy Studies, Chicago Council on Global Affairs, and Korea Economic Institute at the end of 2012 showed that just over half of South Koreans viewed China as a partner (53.5 percent) rather than a rival (46.5 percent), with an overwhelming 94 percent expressing support for the US-ROK alliance." Which means that South Korea sees the US as a more important partner than China. A compilation of South Korean data from 1997-2012 shows that from 2006 to 2011 favorable views on China in South Korea declined from 73 percent to 53 percent.

Relations were further strained when, in July 2012, China released four South Korean activists who had been detained because they assisted defecting North Koreans and supposedly endangered China's "national security." This action sparked controversy as one of the detained activists, Kim Young-hwan, claimed that Chinese authorities tortured her. South Korea's position continues to grow more difficult, as Rozman writes: "While US-China identity conflicts became more salient with the increase in regional tensions from 2009, South Korea's reliance on China and the United States for growth and security makes a stable US-China relationship a top priority." This puts South Korea in the place of mediator that could help bring the two powers to the table. Koh Choong-suk, president of South Korea's Society of Ieodo Research, said in August 2011 that "Asia's mistrust of China and fear of Beijing is based on its territorial ambition ... China's ambition should be counterbalanced by the United States as a Pacific partner to Asia-Pacific nations." That same month, China conducted aircraft carrier trials. In response, a Korea National Defense University professor said, "Korea can secure military deterrence by reinforcing joint deterrence capacity with the United States."[95] While South Korea can become a

---

95 Rozman, Gilbert (an American sociologist specializing in Asian studies). Joint U.S.-Korea Academic Studies: East Asian Leaders' Geopolitical Frameworks, New National Identity Impact, And Rising Economic Concerns With China 2020|Vol.31. :kei_jointus-korea_2020_final_digital_0.pdf (keia.org). Published

mediator between America and China, the South Korean government wants to reinforce their ties with America to create a deterrence to China, and then to sit down and talk with both sides.

As South Korea is partnered with two of the strongest countries in the world, its government must move carefully to prevent the nation from being dragged into war. South Korea wants to strike a balance between their economic relations with China and their military relations with the United States. These partnerships have their limitations, as South Korea has grievances concerning both relationships. South Korea's significance in the South China Sea may allow it to become a mediator between China and America and prevent a major war from breaking out in the region.

by Korea Economic Institute of America, Washington, D.C., U.S.A. 2020. 98-100, 105-107.

Chapter 7: India: A Nation Sharpening Its Blade

Ever since India gained independence from Great Britain, it has become one of the major powers in Asia. Their relationship with China has remained shaky after the 1962 Sino-Indian War, in which China invaded India for northwestern territory that they claimed. This resulted in India viewing China as a regional rival, and also modernizing its military with some equipment manufactured internally and some acquired from financial partners. Much like America and Japan, India has a vested interest in containing China and perhaps creating a buffer zone between India and China by encouraging the independence movement in Tibet.

Although India may not be directly involved in the conflicts of the South China Sea, it has a strong interest in the Malacca Strait, which functions as a naval bridge to the region. Nandini Jawli reports on how vital the strait is to India, and how this affects relations with China. Jawli states that 55 percent of India's trade passes through the South China Sea, which makes the region essential to maintaining the country's economy. The Indian external minister of the state, General VK Singh explained in the Upper House of Parliament on August 4, 2016, that India supports the peaceful resolution to disputes, and that states in the South China Sea should restrain themselves from using force.

As far back as 2000, the Indian Navy first deployed in the South China Sea. As Indian commentators described it "for the first time, in a quiet show of strategic reach, India is going out of her own sphere, experimenting with something new." Jawli states that on May 18, 2016, four ships of the Indian Navy were deployed for a two-and-a-half-month operation in the South China Sea and the Northwestern Pacific. This operation also helped with diplomatic ties, since the deployment made port calls to Vietnam, the Philippines, Japan, Malaysia, South Korea, and Russia. India has conducted bilateral naval exercises with Singapore, and also had friendly port calls with Brunei, Indonesia, Malaysia, the Philippines, and Vietnam. India also has done MILAN (Hindi for "meeting") multilateral naval exercises at Andaman and Nicobar Islands since 1995, in cooperation with Thailand, Malaysia, Singapore, Indonesia, Brunei, Vietnam, and the Philippines. In 2014, the six-day MILAN exercise included India, Australia, Bangladesh, Cambodia, Indonesia, Kenya, Malaysia, Maldives, Mauritius, Myanmar, New Zealand, the Philippines, Seychelles, Singapore, Sri Lanka, Tanzania, and Thailand.[96] India's moves in the South China Sea are those of a nation that is exploring new geopolitical territory to protect its economy.

India's need for a strong navy is a result of tension with China. The cause-and-effect relationship between China's aggression and the strengthening of India's navy is explored by Sudarshan Shrikhande, who writes that India needs to work on leveraging its maritime power against China. He explains that the skirmish in mid-June 2020 in the Sino-Indian border in Ladakh showed why it is important for India to focus on China's strategic intent. India's defense will depend on its ability to overcome Chinese maritime supremacy in the Western Pacific. Both India and China have expanded their militaries since the last war they had in 1962. Shrikhande suggests India should use "sea denial" on China, which is the economic strangulation of a nation and prevention of logistics to get across the trade routes. This is what Germany did in WWII with its U-boat campaigns in the Northern Atlantic.

Shrikhande says that "trade is China's jugular." He also recommends that QUAD become a multi-domain militarily arrangement to counter China's military. He then cites a recent media report that a high-level committee advised the Indian government to "Use [the] Indian

96 Jawli. 89-91.

Navy to Pile Pressure on China." One of the options this committee proposed was to "aggressively impose sea denial in the Indian Ocean." The media rephrased it as "threaten Chinese shipping in the Indian Ocean." Shrikhande mentions however, that naval budget allocations have declined in recent years.[97] The Indian Navy can prove to be useful for the QUAD, but only if they are adequately financed.

The Sino-Indian relationship has been rocky since the late 1950s. Sangit Ragi starts by claiming that the border tension between India and China has existed since the independence of India in 1947. He explains that India wanted to keep the same borders that the British Raj had with China, but China was reluctant to give in to Indian claims. When China was building their Sinkiang-Tibet Road in 1951, the Indians were unaware of the activity until they found it on maps in 1958. India decided to downplay the issue and not make a political protest. As a response to this on September 10, 1959, then Prime Minister of India Rajya Sabha said: "… we have always looked upon the Ladakh area in northern India as a different area as, if I may say so, some vaguer area so far as the frontier is concerned because the exact line of the frontier is not at all clear as in the case of the MacMahon Line. When we discovered in 1958, more than a year ago, that a road had been built across Yehchong in the north-east corner of Ladakh, we were worried. We did not know where it was. Hon. Members asked, 'Why did you not know before?' It is a relevant question, but the fact of the matter is that we just are not within a hundred miles of that area. It is an uninhabitable area, and it has not been under any kind of administration. Nobody was present there. It is a territory where not even a blade of grass grows, about 17,000 feet high."

In 1961, China began to annex parts of the Ladakh region. On

97 Shrikhande, Sudarshan (A retired Indian Admiral, a graduate of both the Soviet Naval War College (1988) and the U.S. Naval War College (2003). He has served at sea and been defense attaché in Australia). "Harnessing Indian Sea Power Post-Galwan: Considerations of Time, Space and Force." :ORF-Issue-rief-379-Sea-power.pdf (orfonline.org). *ORF Issue Brief*, July 2020, Issue No. 379. Published by Observer Research Foundation. New Delhi, India. July 10, 2020. 1-2, 4, 6, 9-10.

July 5, 1962, an Indian military patrol established a patrol post at the Galwan Valley region. The following week, the post was surrounded by 70 Chinese soldiers who were soon reinforced to become a battalion. Between the 15th and 18th of July the Indian Army tried to reinforce the post but were stopped by the Chinese. The siege ended on October 20, 1962, when the Chinese troops attacked on all sides and overran the post in a matter of hours. Prime Minister Jawaharlal Nehru acknowledged this failure saying, "We were getting out of touch with reality in the modern world and we were living in an artificial atmosphere of our creation. We have been shocked out of it, all of us, whether it is the government or the people; some have felt it less, some more." This loss in the Sino-Indian War led India to lose territory, but taught important lessons, such as the need to reorganize and restructure the army, the need for specialized mountain divisions, and the importance of expanding the Indo-Tibetan Border Police. All of this would need the support of a modernized weapons system.

Clashes in the Galwan Valley persist into the present, but India never faced another defeat as they did in the Sino-Indian War. In recent years, India has been building strategic relations with the United States, Russia, France, and Israel to get the most modern weapons, including S-400, Rafael, Apache, Chinook, Globemaster, and Hercules. If those recent acquisitions are any indication, Raji contends, then India is on the right path toward building its strength through important strategic partnerships in time to face challenges from China. The standoff in Ladakh is a test of will in which India has dared to stand up against China's bullying tactics. He adds that China's aggression was what made India pursue the QUAD again. Ragi cites Indian diplomat Jayant Prasad saying, "Beijing's belligerence has been forcing India for a tougher position."[98] India's remembrance of its defeat by the Chinese during the Sino-Indian War and China's continuous engagement on India's border, have pushed the nation to prepare for another war with China.

---

98 Ragi. 3-6, 9-10, 12, 14, 15, 39.

Disputes over the border at Ladakh have affected Sino-Indian relations as recently as 2020. The border clash and its aftermath are covered in detail by Arzan Tarapore, who explains that in May of 2020, China commenced several incursions across the Line of Actual Control (LAC) in Ladakh. Both India and China reinforced their border forces with tens of thousands of troops. Tarapore elaborated that the Ladakh crisis poses a challenge to India and their relationship with China. He contends that, because of the crisis, India now lives in the reality of China being an adversary, with their border being militarized and violence-prone. India also faces the difficult choice of modernizing the army to protect the border with China or modernizing the navy and protect their interests in the Indo-Pacific.

Meanwhile, China has a better-resourced military that can absorb the material cost of mobilization. Tarapore quotes Yogesh Joshi, then commander of the Indian Army's Northern Command, who said during the Ladakh crisis that "we were absolutely on the brink of war." He then cites External Affairs Minister S. Jaishankar's comment that the crisis "profoundly disturbed the relationship (between India and China)." This is reinforced by former Indian ambassador to China and National Security Adviser Shivshankar Menon, who said that "political relations will now be more adversarial, antagonistic, and contentious … there is no grand bargain, no modus vivendi, no accepted rules of the road to be had."

Tarapore writes: "In total, some 40,000–50,000 Indian soldiers were deployed to eastern Ladakh. These troops were joined by India's most advanced main battle tanks, self-propelled howitzers, surface-to-air missiles, and brand-new Rafale multirole fighters." He explains that 22 Indian divisions face Pakistan, while 14 divisions face China, with two divisions in reserve. Tarapore cites a 2013 edition of *The Science of Military Strategy* which had the latest version of the PLA's strategic text. It said that China saw India as a rising power that would impede China's quest to control the "two oceans region" or the Indo-Pacific. The text added that "(India's) strategic deployments will be reflected more in its intentions to control the South Asian subcontinent and the Indian Ocean, and it will give more stress to paying attention to both the land and the sea."

Tarapore comments that "China is better positioned to weather the operational and strategic effects. No doubt the PLA's new deployments and supporting infrastructure have been costly. But the relative cost of this increased militarization—its impost on the defense budget—is likely to be more manageable in China, because China's defense budget is three to four times larger than India's." He further explains that the PLA Western Theatre Command already has 200,000 troops, so it is unlikely that China will shift any more resources. The Ladakh crisis opened the potential of disrupting the Sino-Indian trade relationship and deepening India's relationship with America. Foreign Minister Jaishankar acknowledged that "the India-China relationship is today truly at a crossroads." Reportedly, America provided intelligence support and emergency supplies like extreme-weather clothing to the Indian army reinforcements in the LAC, along with temporary access to two Sea Guardian drones, which provide long-endurance surveillance over a wide area. In November 2020, India agreed to Australia's inclusion in the Malabar exercise. On March 2021, the first QUAD national leaders' summit was held. This partnership with America could give India the confidence needed to build up the military infrastructure on the Line of Actual Control (LAC) in Ladakh.[99] The clash over Ladakh has hampered the relationship between India and China and has brought China's biggest rivals together.

India's push to get involved in the South China Sea has coincided with the tightening of relations with major and minor powers in the region. Utkarsha Mahajan explains that India historically has been wary of joining either a China-led or Western-led coalition in the Indo-Pacific.

---

99 Tarapore, Arzan (Fellow at NBR, adjunct at RAND. Former Australian Defence Department). "The Crisis after the Crisis: How Ladakh will Shape India's Competition with China.": TARAPORE Crisis after the Crisis PDF FINAL 05_05_2021(2).pdf (lowyinstitute.org).

However, in 2015, India had dialogue with the United States and Japan concerning maritime security in the Indo-Pacific. That same year, India collaborated with Japan on infrastructure project. Japan announced a Partnership for Quality Infrastructure initiative to promote quality and financially responsible infrastructure over the "cheap, but shoddy" method that the Chinese employ in their projects. India and Japan have worked together on infrastructure projects in Afghanistan, Bangladesh, Indonesia, Kenya, and Vietnam. To counter Chinese influence, India has also collaborated on projects in Brunei, the Philippines, Vietnam and Singapore.[100] For nearly 10 years, India has been working with the nations of the South China Sea to contain China.

Though it is a given that a nation as big as India would want some influence in one of the most profitable regions in the world, it is odd how the nation has limited involvement in the South China Sea up until now. India's increasing interest into the South China Sea is brought up by Abhijit Singh who claims that India's growing interest with the South China Sea is due to three factors:

1) Trade between India and the nations of the South China Sea has increased threefold, from 21 billion US dollars from 2005 to 2006, to 65 billion US dollars from 2015 to 2016.
2) India wants to uphold international maritime law as it applies to the South China Sea.
3) India is interested in cooperating with its neighbors to contain the People's Liberation Army Navy (PLAN).

---

100 Mahajan, Utkarsha (Doctoral candidate at the Department of geopolitics and international relations, Manipal academy of higher education. Previously she has completed her Post graduation [M.A.] in Politics and Bachelor of Engineering in electronics and telecommunication from University of Mumbai along with two PG Diplomas in International and Security Studies.). India's Engagement In the Indo-pacific: A National Security Perspective.:INDIAS-ENGAGEMENT-IN-THE-INDO-PACIFIC-A-NATIONAL-SECURITY-PERSPECTIVE-Revised-as-of-2nd-Feb-2021.pdf (finsindia.org), Published by the Forum of integrated National Security (FINS). Dadar, India. April 28, 2021. 1, 5-8, 12-13.

Singh goes back to 2016, when India participated in naval exercises themed as "humanitarian mine action," and "peacekeeping operations" with 10 members of ASEAN. Chinese naval watchers have been most concerned over Indian exercises with the US, Australia, and Japan. China has objected to India's deepening ties with the US.[101] India's continuous involvement with China's rivals is intended to assure two things: the economic security of India and the maintenance of the status quo through military deterrence.

While India is making strides in diplomacy with the nations in the South China Sea, it would face challenges in a time of war. These challenges are covered by Darshana M. Baruah, who states that China's development of Himalayan roads contrasts with India's neglect of its own roads across the mountain range. This neglect would make it hard for India to move troops and logistics to and from the Himalayas. China also has submarines docked in Sri Lanka and Pakistan. However, unlike the issue involving territory, the Modi administration has a solution for this: build a port at Chabahar, Iran.[102] Though India is confident in its ability at sea, with only 15 percent of the military budget allocated to the navy, it is safe to say that the Indian Navy will have to work with limited materials. This is indicative of India's concern about going to war with China. They instead want to strike a balance in their relationships between the West and China.[103] To be effective in fighting

---

101 Singh, Abhijit (A former naval officer, Abhijit Singh, Senior Fellow, heads the Maritime Policy Initiative at ORF), "The Nautical Dimension Of India's 'Act East' Policy.":PR180409_The-Nautical-Dimension-of-Indias-Act-East-Policy.pdf (rsis.edu.sg), Policy Report April 2018. Published by S. Rajaratnam School of International Studies and Nanyang Technological School in Singapore, 2018. 4-6, 9.

102 Baruah, Darshana M. (fellow with the South Asia Program at the Carnegie Endowment for International Peace where she directs the Indian Ocean Initiative). "India's Answer to the Belt and Road: A Road Map for South Asia":resrep20723.pdf (jstor.org), Carnegie India Working Paper August 2018, Published by Carnegie India. New Delhi, India. 2018. 20-22.

103 Baruah, Darshana M. (a fellow with the South Asia Program at the Carnegie Endowment for International Peace where she directs the Indian Ocean Initiative). "India in the Indo-Pacific: New Delhi's Theater of Opportunity":Baruah_UnderstandingIndia_final1.pdf (carnegieendowment.org), Carnegie Endowment

China in a war, India will have to improve its navy and its logistics. Decisive diplomacy also will be required when it comes to countering Chinese aggression.

While India will have to focus on China in the scenario of war, they will also face a second front with a country against which they have a longer rivalry with than China: Pakistan. Christopher Clary and Vipin Narang discuss recent India-Pakistan relations, pointing out that India has lost 30,000 civilian and military personnel to terrorist and militant violence in the past 30 years. This has resulted in India getting nuclear weapons to take on Pakistan. Since the late 1980s, India has struggled to deal with Pakistan-sponsored terrorism in the wake of Pakistan acquiring nuclear weapons. Tensions over the Kashmir region led India to consider "develop[ing] the ability to conduct limited war that stopped short of triggering Pakistan's use of nuclear weapons." In May 1998, India began to test its own nuclear weapons. Then Indian Prime Minister Atal Vajpayee argued that India needed nuclear weapons considering Pakistan and China have nuclear arsenals of their own. Pakistan then tested their own nuclear weapons that same month. Clary and Narang write: "A much publicized visit by Prime Minister Vajpayee to Pakistan in February 1999 led some Indian officials to conclude that the two countries' overt nuclear capabilities had 'eliminated the threat of another war, making a settlement possible.'"

However, there are problems brought up by the authors that make peace between India and Pakistan difficult. In 2011, Pakistan conducted another nuclear weapons test, symbolizing the continued cold war between the two nations. India has been developing nuclear weapons and countervailing forces to address Pakistan's long-range nuclear systems. Clary and Narang also mention that India could employ any of three strategies in a war against Pakistan. One to use nuclear counterforce targeting that would give India "a limited ability

---

For International Peace Working Paper June 2020, Published by Carnegie Endowment For International Peace. Washington, D.C., U.S.A. 2020. 5-18.

to disarm Pakistan of strategic nuclear weapons." However, this would lead to the death of millions of Pakistani civilians. The authors suggest that India could take a page out of America's book and maintain limited nuclear options to incentivize Pakistan into taking a disarming strike instead of a nuclear strike. Currently, India has been preparing a strategy to disarm Pakistan of its battlefield nuclear systems. This is similar to how President Reagan bluffed about military development and made the Soviet Union back out of the Arms Race.[104] The two-front war that India could face against Pakistan and China is a scenario that India and its partners will have to plan for in the case of a South China Sea War.

Much like its rivals, India is making efforts to modernize their military. As Romana Fahmeed, Shahid Bukhari, and Shakeel Ahmad collaborated on describing this effort, they stated that India is working on a modernization campaign to fill in any gaps in its military and maintain its claim as a great power. In recent years, India has lacked the economic prosperity to subsidize the cost of military modernization. Fahmeed et al write: "With the second-largest population in the world, while having a significant size of territory and economic strength, it is natural for India to desire for having a great power status in world politics." India is the world's largest importer of weapons and has signed 169 defense deals for a period of 15 years (2012-2027). In 2019, India's defense budget was 49.68 billion American dollars. From 2020-2021, that increased to $73.65 billion. However, Indian analysts are dissatisfied with the allocation of said funds and want 90 percent of them allocated to current inequities with the military. The authors cite

---

104 Clary, Christopher (Christopher Clary is an Associate Professor of Political Science at the University of Albany.), Narang, Vipin (Vipin Narang is the Principal Deputy Assistant Secretary of Defense for Space Policy. Narang has been an American political scientist. He is Professor of Political Science at MIT.). "India's Counterforce Temptations: Strategic Dilemmas, Doctrine, and Capabilities" :India's Counterforce Temptations: Strategic Dilemmas, Doctrine, and Capabilities | International Security | MIT Press. *International Security* (2019) Volume 43 Issue 3: 7–52. Published by MIT Press. Cambridge, Massachusetts, U.S.A. February 1, 2019. 7-9, 11-12, 15-16.

a report by a defense news site saying: "due to the shortage of funds, at least a dozen pending defense contracts will experience delays."[105] If India keeps the military properly funded, then they can match China's strides toward modernization.

These seemingly contradictory decisions by the Indian government have much to do with popular sentiment. Dhruva Jaishankar delves into this by citing a December 2018 survey from India's strategic community in which 54 percent see China's aggression as the biggest external challenge; 2 percent think India should collaborate more with China; 75 percent see America as the most important partner on global issues; 43 percent believe that India should collaborate more with America; and 74 percent see the Indo-Pacific as the important framework for India's extended neighborhood. Majorities identify the China-Pakistan Economic Corridor (73 percent) and Chinese investment in Sri Lanka and the Maldives (63 percent) as major security challenges for India. Jaishankar states that even though most Indians see America as their country's most important partner on global issues, 54 percent want their country to remain neutral between America and China. The survey showed that 73 percent view BRICS (Brazil, Russia, India, China, and South Africa) as an important organization for India's interests. However, 54 percent of Indian elites and 46 percent of the general public believed that China posed the biggest external problem for India.[106] As India moves to counter China,

---

105 Fahmeed, Romana (Research Scholar, Department of Political Science, Bahauddin Zakariya University, Multan, Pakistan), Bukhari, Shahid (Associate Professor, Department of International Relations, Bahauddin Zakariya University, Multan, Pakistan), Ahmad, Shakeel (Research Fellow, Institute of Air and Space Law, McGill University, Canada). "India's Military Procurement Programs & Economic Capacity: Compatibility & Pragmatism." :India's Military Procurement Programs & Economic Capacity: Compatibility & Pragmatism | Journal of Business and Social Review in Emerging Economies (globalcsrc.org). *Journal of Business and Social Review in Emerging Economies* Vol. 6 No 1 (2020). Published by CSRC Publishing, Punjab, Pakistan. March 31, 2020. 135-136, 142.

106 Jaishankar, Dhruva (Executive Director of the Observer Research Foundation America [ORF America], which he helped establish in 2020. He is also a Non-Resident Fellow with the Lowy Institute in Australia and writes a regular column

it will be important for the government and the people of the country to be on the same page for a healthier democracy.

India's relationship with Tibet and Taiwan also affects its relationship with China, as China seeks to solidify control over these regions. David Scott covers India's complex relationship with China. Scott claims that in the wake of the Galwan (Ladakh) Crisis, some in India are considering "rethinking" the One China Policy in terms of Tibet and Taiwan. He explains that while the Himalayan border region, including Tibet, can become the subject of a "great-game" competition (where two countries fight for regional influence), it is probably too late for India to support Tibetan independence. He notes that, in 2003, India recognized that the "Tibetan Autonomous Region is part of the territory of the PRC."

India could join in calls for a genuine Tibetan autonomy instead of the Chinese Tibetan Autonomous Region (TAR), and India could use Chinese human rights abuses in Tibet as further justification for said autonomy. Scott goes on to describe that despite India saying in a 2003 joint statement that "it does not allow Tibetans to engage in anti-China political activities in India," this point could be reconsidered through support and cooperation with the India-based Central Tibetan Administration (CTA). India can deploy the Special Frontier Force (a force that recruits ethnic Tibetans) to the Chinese border. Finally, the Indian government could make more contact with the Dalai Lama. Scott proclaims that there has also been a call to focus on Taiwan. Indian analysts see a match with India's Act East policy and Taiwan's Go South policy. This would anger China, but as Scott writes, "if it (China) is not going to respect India's sensitivities (anyway), there is no reason why New Delhi should be unduly deferential towards Beijing's interests."

He goes on to discuss India's strategic relationships with Vietnam,

for the Hindustan Times). "Survey of India's Strategic Community." :Survey of India's Strategic Community (think-asia.org). Impact Series | March 2019. Published by Brookings Institution India Center. New Delhi, India. 2019. 3, 10, 13, 17, 21, 24.

South Korea, and Mongolia. The Indo-Vietnamese relationship has strengthened since a Defense Protocol was signed in 2000, with the Joint Vision being signed in December 2020. South Korea wants to avoid being dependent on China, thus they have been cooperating in an economic and defense sense with India. Mongolia, wanting to avoid being absorbed into China, signed a defense cooperation agreement with India in 2019. Scott then goes into India's relationship with the major powers in the South China Sea, He explains that India has joined with Japan and America in the India-Japan-US (IJUS) trilateral cooperation and with Australia and France in the Australia-France-India (AFI) trilateral mechanism. India also had a five-way exercise with America, Japan, Australia, and France after the battle of the Galwan Valley. Although India is also involved in the Russia-India-China (RIC) trilateral arrangement since 2001, that has nothing to do with Indo-Pacific cooperation nor with mediating ties between India and China. Similarly, the BRICS group does not address Indo-Pacific cooperation nor any regional policy differences between India and China.

COVID-19, and its cover-up by the Chinese government, further deepened India's distrust of China. Scott cites the words of an Indian commentator in mid-2020: "China is now a complete threat to India, as traditional (military) and non-traditional (health) hazards are visible in the current situation." He then cites S. Jaishankar, a diplomat and minister of external affairs, saying "peace and tranquility in the border areas is the basis for development of relations in other domains. If they are disturbed (by China), so inevitably will the rest of the relationship."[107] India's connection to China, while significant,

107 Scott, David (a member of the Center for International Maritime Security). "India's China Challenge: Foreign Policy Dilemmas Post-Galwan and Post-Covid." :India's China Challenge: Foreign Policy Dilemmas Post-Galwan and Post-Covid | The Journal of Indian and Asian Studies (worldscientific.com). *The Journal of Indian and Asian Studies* Vol. 02, No. 02, (2021). Published by World Scientific Publishing Co Pte Ltd. Chennai, India. 2021. 5-11, 13, 16.

is hampered by decades of troubling history between the two major powers.

While India and Pakistan have contentions over the Kashmir, they also have butted heads over Balochistan, a region that includes part of Pakistan and also borders India, Afghanistan and Iran. This is examined by Prasanta Sahoo who claims that, in light of Pakistan supporting anti-Indian terrorists in Kashmir, India has started to support terrorism in Balochistan. India has also used its diplomatic power to choke Pakistan's economy and create political instability throughout Pakistan, according to Sahoo. He examines how India's Balochistan tactics contributed to the shattering of Pakistan's dream in Kashmir. In August 2016, during an Independence Day Speech, Modi denounced the Pakistani Army's atrocities against the people of Balochistan, Gilgit, and Pakistan-occupied Kashmir. Strategists and academics have studied this change in national security strategy, and all neighboring countries support Modi's stance, according to Sahoo: "Modi's 'zero tolerance' policy on terrorism forced the government to give a free hand to the army to punish the perpetrators across the border. At midnight on 29 September 2016, special commandos of the Indian army launched their first ever surgical strikes across the line of control (LoC) in Pakistan occupied Kashmir (PoK). After the raid, which left Pakistan stunned, the rest of the world appreciated India's daring move to avenge the assault on its sovereignty and territorial integrity."

Sahoo goes into another example: After a 2019 attack on Indian army vehicles in Pulwama led to the death of 43 soldiers, India responded with airstrikes on the headquarters of Jaish-e-Mohammed (JeM), a Pakistan-based terrorist organization. Afghanistan, Bangladesh, and the Maldives have backed India's position, and denounced Pakistan's interference in Jammu and Kashmir and its involvement in the attacks at Uri and Pulwama. Major-scale violence in these contested areas has started since Modi's anti-terrorist stance. Sahoo explains that Balochistan or "Pakistan's dark underbelly," is rich in natural resources and thus is a region sought by all. He then recounts that in

2006, Baloch national leader Nawab Akbar Khan Bugti, was killed by Pakistani fighter planes.

Sahoo shows that this was not an exception to the rule, as atrocities by the Pakistani Army against the Baloch people include murder, the torture of young boys, kidnappings, rape, and the forced conversion of young girls. Sahoo goes further by writing: "Baloch activists like Naela Quadri Baloch and Tarek Fatah have reported a long list of crimes committed by Pakistan against the Baloch people. Nearly 25,000 people, including men, women and children, have been abducted and murdered. As per the BBC and the Asian Human Rights Commission, on 25 January 2014, hundreds of mass graves were found in Balochistan. ... Balochistan has been pleading with the United Nations to intervene through its 2012 declaration against genocide."[108] Both Kashmir and Balochistan will become frontline regions in the scenario of a war between Pakistan and India; they could also become battlefields in the South China Sea War.

While the United States benefits from its partnership with India on the South China Sea issue, other US partners, including Japan, have also gotten closer to India. The strengthening of relations between Japan and India is covered by Satoru Nagao, who explains that the India–Japan strategic partnership ties Indian Prime Minister Narendra Modi's "Act East" policy with Japanese Prime Minister Shinzo Abe's push for a "Free and Open Indo-Pacific." He cites Modi, saying that no other strategic partnership "can exercise a more profound influence on shaping the course of Asia and our interlinked ocean regions." Nagao highlights parts of the India-Japan strategic partnership, such as Japan joining the annual Malabar naval exercises with India and America as a regular participant in 2015. He then asserts that as the US moved to

---

108 Sahoo, Prasanta (PhD Advanced International Relations Theory). "India's Balochistan Tactic: HAS IT SHATTERED PAKISTAN'S KASHMIR DREAM?.":India's Balochistan Tactic on JSTOR. *World Affairs: The Journal of International Issues*, Vol. 23, No. 2 (SUMMER 2019 [APRIL-JUNE]), pp. 98-115. Published by World Affairs. New Delhi, India. June 2019. 98-102, 108-111.

forge better ties with India at the turn of the 21st century, Japan could not afford to be left behind.

China's increasing pressure on the South China Sea has led to deepening ties among the US, Japan, and India. Nagao continues by claiming that, despite the rivalry between China and America, both states are economically intertwined and could not attack each other without the threat of a nuclear war. Japan has economic interdependence with both America and China. And India and China are militarily interdependent as nuclear powers. In this complex situation, strategic partnerships offer more benefits than traditional organizations like NATO, which is why India, Japan, and America are working in a strategic partnership rather than a traditional alliance, according to Nagao. "Unlike alliances … (strategic partnerships) are not expressly aimed at an identified enemy. This leaves space for building bridges with strategic adversaries with whom conflict is not a viable option owing to interdependence. Strategic partnerships are also limited and flexible and do not involve deep action-inducing commitments to the security or even political positions of partners."

Strategic partnerships help nations collaborate in the realms of security and economy. During a visit to India in 2015, Japanese Prime Minister Abe said that Japan and India had the "bilateral relationship with the greatest potential in the world." Nagao writes that there is "cautious optimism" for this partnership, since there are differences in worldview that can form a barrier for deepening relationships for the two countries. He elaborates that Japanese Official Developmental Assistance (ODA) to India in 2014-2015 is valued at $15.5 billion. The success of the Japanese-funded Delhi Metro Project demonstrated the impact that Japanese money and technology have on Indian infrastructure. In December 2015, India agreed to purchase a high-speed bullet train for the Ahmedabad-Mumbai route from Japan worth $15 billion in concessions and loans. Japan is also supporting Modi's "Make in India" initiative through a $12 billion fund.

Former Prime Minister Manmohan Singh observed that "India

needs Japanese technology and investment," and that India offers opportunities "for the growth and globalization of Japanese companies for the overall prosperity and growth of Japan." In 2013, the first Japanese National Security Strategy "recognized India along with China as the 'primary drivers' of change in the international 'balance of power.'" Nagao explains neither Japan nor India want a China-led security order. By 2014, China's economy was twice as big as Japan's economy, and five times the size of India's, though India is expected to become the fastest-growing economy by the next decade. China spends twice as much on its military as India and Japan combined. India is also concerned about China's construction of artificial islands. While China is wary of the close relationship of India and Japan, India is more focused on border clashes with China in the Himalayas, and Japan is focused on maritime concerns. However, both India and Japan are concerned about the possibility of the United States and China forming a stronger relationship. India was scared of this since their nuclear tests in 1998, after which China and America issued a statement "to 'jointly and individually' work towards peace in South Asia." Japan has had a similar fear since 1971, when America decided to normalize relations with China without saying anything to Japan. Nagao writes: "For India, Japan has (also) emerged as an indispensable partner in its attempts at economic modernization and is integral to its efforts at preserving a favorable balance of power in the Indo-Pacific. For Japan, the importance of its strategic partnership with India is surpassed only by its alliance with the United States and its 'sub-alliance' with Australia."

When Abe visited India in late 2015, a former Indian ambassador to Japan commented that the concluding joint statement was "remarkable for its strategic resolve, clarity of purpose and joint actions … It marks a qualitative new phase of the India–Japan strategic partnership with vast region-wide ramifications." However, Nagao notes that there is skepticism about the limitations of this strategic partnership. One example was Japan and India's responses to the Nepal earthquake in

2015. Though there were great contributions from both countries, neither cooperated with the other. There have also been difficulties in cooperating in the private sector. Nagao writes: "The governments of Abe and Modi have found ways to transcend or overcome many of these difficulties, whether in forcing an agreement on civilian nuclear cooperation, signing a high-speed railway agreement, or initiating the trilateral dialogue with Australia. Japanese defense companies are inexperienced compared to the French and German competitors. Japan and India should focus on other rivals such as Pakistan and North Korea. India has been dealing with resource constraints, with the navy suffering the most from it; being summarized as 'frequently been stymied by a series of infrastructural, organizational and political failings.'"

Nagao then cites Japanese defense minister Itsunori Onodera who said that a defense relationship is lacking among the countries in the South China Sea. For example, America, Japan, and Australia are allies yet have no close security ties with each other. India is restrained in the use of its military. Aside from peacekeeping operations, the Indian military has not crossed its own borders since 1972. A main reason for India's restraint is that India would not gain much from attacking their smaller neighbors to reciprocate. India prefers rather to encourage cooperation with those neighbors. India also cooperates on the international scene. However, there have been 60 joint US-India exercises over the 2010s, Japan and India had seven joint exercises. Nagao claims that Japan will provide enough force against China that India would not have to deal with the full force of the Chinese military. While Modi was visiting the Rose Garden in the White House in June 26, 2017, then-President Trump said "(O)ur militaries (The US and India) are working every day to enhance cooperation between our military forces. And next month, they will join with the Japanese navy to take part in the largest maritime exercise ever conducted in the vast Indian Ocean."

The ties among America, India, and Japan are growing. As Nagao writes: "The fact that both India and Japan are democracies also makes it easier for the United States to advocate and pursue closer ties with both countries in terms of domestic politics. American domestic constraints to improving ties with New Delhi and Tokyo are insignificant compared to constraints to improving ties with other major countries such as China and Russia ... (T)he United States regards both Japan and India as stakeholders in the international liberal order, not challengers."[109] India now has greater ties with China's other rivals as they become a part of a coalition to contain China.

With Chinese nationalism on the rise, Hindu nationalism has been growing in India as years go by. Gilbert Rozman elaborates that since independence, India has been at the crossroads of two nationalisms: Indian nationalism, which focuses on a more secular view of India along with its diverse ethnicities, languages, and cultures; and Hindu nationalism, which sees India as a nation that can be ruled by Hindus after hundreds of years of being ruled by Muslims and the British. To founding fathers of Hindu nationalism like Vinayak Damodar Savarkar, Indian territory is not only a fatherland, but holy land to Hindus, Buddhists, Jains, and Sikhs. Rozman writes: "For most foundational Hindu nationalist authors, Hinduism's value was defined in contrast to the West ... For example, Modi's favorite Hindu nationalist thinker, Deendayal Upadhyaya, put forward a concept he called 'integral humanism' that outlined a vision of Hinduism's contributions relative to Western thinking." Hindu nationalists see Buddhist Tibet as a part of the larger Hindu family, given that Buddhism started in India and there are several Indian holy sites in Tibet.

Rozman claims that ever since the victory of Modi's Bharatiya Janata Party (BJP) in May 2019, Hindu nationalism has played a "potent" role in reshaping the country. But it must be said that Hindu nationalists

---

109 Nagao, Satoru (a fellow [non-resident] at Hudson Institute, based in Tokyo, Japan). "Why Japan Needs India as a Defence Partner." India and Japan: Assessing the Strategic Partnership. Pp. 65-89. Published by Palgrave Pivot. Singapore. March 21, 2018. V, 2-3, 5-7, 14-15, 25-26, 31-32, 34-36, 55-58, 72, 76, 78-80, 92, 96-97, 99.

focus their concerns more on Islam than on either China or the West. Rather, it has been the Indian elite and bureaucracy that has expressed concerns about China since the Sino-Indian War in 1962. Rozman quotes Chinese media outlet *Global Times* as writing: "Hindu nationalism risks pushing India into war with China." However, in America, there is a more optimistic view to the effect that Hindu nationalism may cause India to move closer to America and further from China. Rozman however, notes that while Hindu Nationalists will criticize China for its policies in Tibet or for dumping products in India, they also criticize the West for their encroachment of their values or westernization.[110] Though Hindu nationalism is centered on the internal affairs of India, holy sites in Tibet could inspire nationalists to call for war with China.

India has significant connections with countries in America's sphere of influence, but also strong connections with neutral nations. Karthik Nachiappan explains how issues such as the economy, counterterrorism, and cybersecurity bring India and ASEAN together. As of August 2018, ASEAN became India's fourth biggest trading partner. He writes: "India's trade with ASEAN now stands at US $81.33 billion, approximately 10.6 percent of India's overall trade. India's exports to ASEAN totaled around 11 percent of her total exports … In 2019, India's top three trading partners in the ASEAN were Indonesia ($21 billion), Singapore ($18 billion), and Malaysia ($15 billion), largely dominated by a range of goods from India and commodities and manufacturing items from these three ASEAN states."

Despite its size, India's trade with ASEAN trails behind China sevenfold. In 2020, China's trade with ASEAN was $700 billion. From 1995 to 2017, trade between China and ASEAN increased 10-fold. Nachiappan notes that India's ties with ASEAN may deepen as a result of shared concerns such as cybersecurity and counterterrorism. In fact, since the early 2000s, India and ASEAN have been engaged in mutual counterterrorism measures. More specifically, India has

110 Nagao. 104-105, 118-119, 128.

given the ASEAN nation the Philippines $500,000 to support their efforts to fight the Islamic State. He states there is room for India to further deepen its cooperation with ASEAN, primarily with individual ASEAN members such as Malaysia, Indonesia, Thailand, and Vietnam. With nearly 660 million internet users, India has a digital economy of $200 billion, which is poised to grow to $800 billion by 2025. ASEAN has 300 million internet users that can collectively generate a digital economy worth $150 billion annually. Nachiappan claims that the digitization of India and ASEAN breeds the need for further cyber-security measures: "With rising cyberattacks, cybercrime, phishing, ransomware, misinformation, and data fraud, there is a need to bridge differences in approach and capabilities and enhance cyber resilience."[111] India's connections with ASEAN could help them and QUAD proper gain an important ally.

India's potential to impact the South China Sea War should not be underestimated as they, along with America, could impose a blockade that hinders China's ability to fight a prolonged conflict. The combination of their own military armor and weaponry with military supplies from the United States, Russia, France, and Israel, gives India a wide array of abilities on land, sea, and air. Given how many Tibetans are in India and how close Tibet is to India, the Tibetans' role in the Indian Army could be pivotal in fighting in their homeland. India's contribution to the QUAD has provided the Indian military a newfound sense of cohesion with the other militaries in the partnership. By opening up a land front on a mostly seafront war, India can divert Chinese resources that would otherwise be used for China's invasion of Taiwan. And India could cooperate with America and other nations in blockading China from getting sea imports.

---

111 Nachiappan, Karthik (a Fellow at the Institute of South Asian Studies, National University of Singapore). "India–ASEAN Relations: Riding and Transcending the 'Indo-Pacific' Wave." :India–ASEAN Relations: Riding and Transcending the "Indo-Pacific" Wave | The Journal of Indian and Asian Studies (worldscientific.com). *The Journal of Indian and Asian Studies* Vol. 02, No. 02, 2140004 (2021). Published by SAGE Publications, New York, New York, U.S.A. 2021. 2-3, 5-6, 10-12.

Chapter 8: Australia: A Reliable Ally or a Reluctant Partner?

Despite its distance from the South China Sea, Australia relies on trade in the region for precious resources. This gives Australia a genuine stake in the region's geopolitics, especially given the growing hostility between the US and China. Like other nations in the South China Sea, Australia has military relations with America, and economic relations with China. The military relationship with the United States dates from World War II. Mineral-rich Australia sends a majority of its ores to China for refining. Given China's ambitions in Australia and the South China Sea in general, Australia has partnered with the US, India and Japan to form the QUAD to contain China.

Much as with all nations in the South China Sea, the politics of Australia is shaped by the standoff between America and China. Russell W. Glenn writes that "Harvard Professor Graham Allison and Chinese President Xi Jinping have cautioned repeatedly on the dangers of China and the United States falling into a 'Thucydides trap' in a historical allusion to the strategic competition between Athens and Sparta 2,500 years ago, which tragically ended a golden age in ancient Greece. Australian Prime Minister Malcolm Turnbull, who is equally fond of this historical analogy, has warned against taking the so-called Asian peace of recent decades for granted." However, Australia still relies on America for help, as Glenn explains: "Australian military presence was arguably more significant in an alliance sense than was any combat success or nation-building outcome they achieved. Australia's military efforts in Iraq and Afghanistan reinforced our reputation as a good global citizen, one committed to a rules-based world order. Yet actions in both countries lacked a strategy effective in integrating Australia's broader government commitments for achieving anything beyond 'showing the flag.'" Both nations are a part of the America-Australia-New Zealand Alliance (ANZUS), which is focused on the economic interests of all parties involved. However, Glenn also notes the possibility of the 'grand bargain' where great powers can coexist, due to an agreement on terms to prevent a war.[112] Australia's close relationship with America and its proximity to China leaves the nation in a precarious situation.

Australia's wartime experience in the War on Terror makes the country a valuable asset in the realm of military. Rhys Crawley elaborates that 25,000 Australians served in Afghanistan from 2001 to 2014, making Australia the largest non-NATO contributor to the International Security Assistance Force (ISAF). The war cost the country 8.3 billion Australian dollars. Australia became involved because of the mutual defense clauses in the ANZUS treaty of 1951, and in response to the 9/11 attacks. Crawley discusses that throughout their time in Afghanistan, the ADF did high-intensity combat, vehicle-mounted reconnaissance, security patrols, direct-action assaults, clearance operations, and civic-action programs. He then cites Dr. Albert Palazzo, director of the Australian Army's War Research Centre, who wrote: "It is only through rapid adaptation that a military organization can keep pace with an adversary who is also evolving. Plainly, both combatants seek an advantage over the other."

Crawley writes: "This selective embed program delivered huge benefits to Australia. Aside from exposing a generation of senior officers

---

112 Glenn, Dr. Russell W. (a graduate of the United States Military Academy). New Directions in Strategic Thinking 2.0: ANU Strategic & Defense Studies Centre's Golden Anniversary Conference Proceedings. :New Directions in Strategic Thinking 2.0: ANU Strategic & Defence Studies Centre's Golden Anniversary Conference Proceedings (oapen.org), Published by Australian National University. Canberra, Australia. 2018. VIII, 30, 39, 43, 54.

to coalition warfare at the operational and strategic levels, it allowed Australia to have a say in shaping the war at the tactical level without having to deploy too many people, expend large sums of money, or put people unnecessarily in harm's way … This reality provides the first lesson for Australia: it must actively implement formal mechanisms to capture the lessons of Afghanistan and adapt at all levels and across all arms of government. It cannot afford to have those lessons confined solely to the realm of informal corporate knowledge, where it risks evaporating as senior public servants and military officers retire."[113] Australia's lessons from the War on Terror could help the country and QUAD in terms of developing a coordinated strategy.

While Australia wants positive relations with both China and the United States in the South China Sea, China's aggression has made this scenario difficult. Going back to Carolyn Bartholomew's and Robin Cleveland's "2021 Report to Congress," in 2020, China began to put economic pressure on Australia via trade barriers. That same year, China issued 14 grievances about Australia. Those grievances were:

1) Since 2018, more than 10 Chinese projects have been rejected by Australia. Putting restrictions on infrastructure, agriculture, and animal husbandry.
2) Banning Huawei Technologies and ZTE from the 5G network over unfounded security concerns.
3) Using foreign interference legislation to target China without any evidence.
4) (The) stigmatization of cooperation between Australia and China and imposed restrictions, including revoking visas for Chinese scholars.

---

113 Crawley, Rhys (a senior lecturer in history at UNSW Canberra, and the author of the Official History of Australian Operations in Afghanistan, 2005-2010). "Australia's Lessons.":"Australia's Lessons" by Rhys Crawley (armywarcollege.edu). The US Army War College Quarterly: *Parameters* Winter 2019 Vol 49 Number 4 Article 8. Published by United States Army War College. Carlisle, Pennsylvania. November 1, 2019. 49, 52-53, 58, 60, 62.

5) A call for an inquiry into COVID-19 as an attack that echoes the American attack on China.
6) The wanton interference in Xinjiang, Hong Kong, and Taiwan affairs, and crusading against China.
7) Australia is the first non-littoral nation to make a statement on the South China Sea to the UN.
8) Siding with America in their anti-China campaign and has joined them in spreading disinformation on COVID-19.
9) Australia aims to end their participation in the BRI.
10) Provided funding to an anti-China think tank, which is peddling lies about Xinjiang.
11) The reckless search and seizure of Chinese journalists' homes and properties without any charges or any explanations.
12) False allegations about cyber-attacks.
13) Condemnation of the Chinese government by MPs and racist attacks against Chinese and Asian people.
14) A media that is poisoning the waters for bilateral relations with unfriendly and antagonistic reporting on China.[114] This shows a growing divide between Australia and China, as both nations prepare for the worst.

In recent years, the Australian government has been more critical of China as Chinese ambitions have grown. This change in relationships between the two governments is explored by Patrick Köllner, who contends that Australia reset its China policies from 2017 to 2018. Australia's security ties with America and economic ties with China puts the nation in a difficult position. In recent years, Australia has emphasized the security issues and been openly critical of China's governance, human rights and foreign policy. Köllner writes, "Australia's greater propensity to openly criticise China was also on display in what John Garnaut (2018) calls the reset of China policy. That reset had internal and external dimensions and led China to

114 Crawley. 52-53.

freeze bilateral high-level political exchange. *The Global Times*, China's international newspaper and propaganda organ known for its blunt language, said in May 2018 that 'Australia's relations with China are among the worst of all Western nations.'"

Köllner goes further to write that "Australian cabinet members publicly adopted an increasingly critical tone towards China in 2017. Speaking at a public forum in Singapore in March, (2017) then foreign minister Julie Bishop made a thinly veiled jab at the PRC, expressing concern about rising powers' use of 'newfound strength to challenge existing territorial or strategic boundaries or to impose their political will on others.' At the Shangri-La security dialogue a few months later, then-PM Malcolm Turnbull (2015–2018) was even more direct, noting fears that China would 'seek to impose a latter-day Monroe Doctrine on this hemisphere in order to dominate the region' and urging the PRC to respect 'the sovereignty of others.'"

Köllner adds, "One of the biggest political scandals in Australia in 2017 saw the fall and eventual resignation of Labor Party senator Sam Dastyari, who had advocated accepting Chinese territorial claims in the South China Sea. He was later found to have accepted substantial donations from a former Chinese billionaire real-estate magnate living in Australia (until his residency permit was eventually revoked in early 2019)." Köllner then explains that by the first half of 2019, China intentionally mishandled Australian coal exports for months on end in order to put economic pressure on Australia.[115] The recent souring of relations means that China's influence on Australia is stunted and gives America the chance to fill that void and increase its influence in the nation.

The United States's relationship with Australia has not only

---

115 Köllner, Patrick (vice president of the GIGA German Institute of Global and Area Studies, director of the GIGA Institute of Asian Studies, and professor of political science [with a focus on Asian studies] at the University of Hamburg). "Australia and New Zealand recalibrate their China policies: convergence and divergence." *The Pacific Review* Vol. 34 issue 3 2021. Published by Routledge at Oxfordshire in Britain. 406. 411-412. 414-415.

strengthened in the past 50 years but has helped Australia open up relations with other nations. John Blaxland claims that for over 75 years, Australia has looked to America for security and toward China as a trade partner. China has overtaken Japan as Australia's greatest trade partner over the past decade. He argues that Australia must strengthen relations with other multilateral and regional powers. Blaxland explains that when America reinvigorated relations with China in 1972, Australia followed that same year with then Prime Minister Gough Whitlam establishing diplomatic relations with China. The relationship has flourished since then. "Today, China is Australia's largest two-way trading partner in goods and services (valued at $150.0 billion in 2015-16), its largest export market ($85.9 billion in 2015-16) and largest source of imports ($64.1 billion in 2015-16)." The Australian Government has diversified this relationship with China. One example is with the 2015 China-Australia Free Trade Agreement. China has also become Australia's highest spending inbound tourism market, with around 1.1 million visits to Australia from Chinese nationals in 2015-16.

Blaxland anticipates that President Trump will uphold the Guam Doctrine of 1969, in which President Nixon said that America will keep up its treaty commitments with the nations in Asia, but said nations in Asia will have to be responsible for their own security. "Rory Medcalf and C. Raja Mohan have called for increased Australian cooperation with India to build regional resilience against the vagaries of US-China relations. They see India and Australia as well placed to form the core of a middle power coalition to build regional resilience and see cooperation possible in a range of areas." While Japan and Australia have relations with America, their exercises in East Timor in 1999, in Aceh from 2004 to 2005 and in Iraq from 2005 to 2006, provided a foundation of cooperation that can help Australia defend against an uncertain future. This was reinforced by Australia helping Japan with relief from the earthquake and tsunami in 2011.

Blaxland further delves into the fact that ever since the end of the Cold War, Australia and ASEAN have enjoyed greater ties in trade, education, tourism, and culture. He writes, "Security ties are important, but trade is a major driver of international relations. The combination of trade ties with ASEAN and the countries of Northeast Asia cumulatively makes the South China Sea transit routes more significant than ever for the future prosperity of many countries including Australia. Ensuring that freedom of navigation has long been an important part of the job for Australia's navy and air force. Historically, however, Australia has studiously avoided taking sides on the numerous territorial disputes affecting the South China Sea."[116] While America has helped Australia when it comes to diplomacy, Australia could also look for diplomatic partners in the region.

Australia tries to maintain relations with China, but China's increasing influence has caused a stir in the country. Amy Searight claims that Australia's ability to balance relations between China and America is hindered by Chinese attempts to influence Australian politics, universities, and Chinese communities living in Australia. Former Australian Prime Minister Malcolm Turnbull introduced a bill to crack down on foreign interference, which was passed with bipartisan support. When the Australian government announced a ban on Huawei and ZTE from their 5G network, China responded with an economic freeze and a slowdown on coal and wine imports. Turnbull went as far as to call it "covert, coercive, or corrupting" and said Chinese actions crossed "the line that separates legitimate influence from unacceptable interference."

According to Searight, China has played a long game in Australia to embed pro-Beijing sentiments in the country. In 2015, Duncan Lewis, director of the Australian Security Intelligence Organization (ASIO),

---

116 Blaxland, John (professor and Head of the Strategic and Defence Studies Centre, in the Coral Bell School of Asia Pacific Affairs, College of Asia and the Pacific at the Australian National University). "Strategic Balancing Act: Australia's Approach to Managing China, the USA and Regional Security Priorities." *Security Challenges*, Vol. 13, No. 1. Published by The Institute for Regional Security in Barton, Australia. 19-20, 22-24, 29.

warned the heads of the three major political parties of Australia—the Australian Labor Party (ALP), the Liberal Party of Australia, and the Nationals—about accepting donations from Chinese donors, for such donations might come with strings attached. Lewis mentions two names: Huang Xiangmo and Chau Chak Wing, both billionaires who had donated to the Liberal Party and the ALP.

Searight provides a lengthy description of the two businessmen: "Dr. Chau Chak Wing is an Australian citizen who gave generously to both the Liberal (AU$2.9 million) and Labor (AU$1.7 million) Parties. He also donated AU$20 million to the University of Technology Sydney (UTS) to build a Frank Gehry-designed business school and AU$15 million to University of Sydney for a new museum. Over the years, Chau cultivated a network of powerful Australian politicians, including former prime ministers John Howard, Kevin Rudd, and Bob Hawke, all of whom visited his palatial estate in Guangzhou, China, where he mostly resides. Chau was clearly well connected in Beijing, given that he helped broker several big deals between China and Australian business, including an AU$25 billion liquefied natural gas export deal between Western Australia and China in 2002 that was heralded as Australia's largest export deal ever …

"Huang Xiangmo was a relative newcomer on the scene, having arrived in Australia in 2011 under somewhat mysterious circumstances, apparently making a 'hurried' exit from China after his close associates and political patrons in his home, the city of Jieyang, were caught up in a major corruption scandal. Huang quickly gained permanent residency and political clout in Australia. He was a major political donor to both the Labor and Liberal Parties, donating more than AU$1 million to each party between 2012 and 2015. Huang's status as 'leader of Australia's Chinese business community' was on display when he was there as guest of the ambassador to greet Xi Jinping on Xi's first official visit to Australia as president in 2014."

Searight goes further to explain that what brought this issue to light was the Dastyari affair. This episode arose from Sen. Sam Dastyari's statement in a pre-election press conference in 2016 organized by Huang that "the South China Sea is China's own affair" and that "as a supporter of China, and as a friend of China," he would help maintain the relationship by "knowing when it is and isn't our place to be involved." Dastyari then added that "Australia should remain neutral and respect China's decision." He made these statements when China's behavior in the South China Sea was a hot-button election issue. Weeks later, Dastyari denied making these statements, but the audio released in 2017 proved otherwise. What's more, Huang paid $5,000 toward his legal bills, and another Chinese donor paid for his travel expenses. Dastyari was forced to apologize and step down from his frontbench party position, due to backlash from Canberra.

In June 2017, the Australian Broadcast Corporation (ABC) aired several explosive allegations following a joint investigation by *Four Corners,* its current affairs program, and *Fairfax Media* newspapers. Among the revelations was a report that the ASIO chief had briefed the major political parties about Huang Xiangmo and Chau Chak Wing, who were suspected of being Beijing's agents of influence. Huang Xiangmo had threatened to cancel a promised AU$400,000 donation to the Labor Party if the party did not soften its stance on the South China Sea. The threat was made in response to Conroy's June 2016 statement supporting freedom of navigation operations. Finally, the *Four Corners/Fairfax* broadcast reported that Sam Dastyari's office had made four approaches to the immigration office on behalf of Huang's citizenship application, including two phone calls by Dastyari himself. The final blow to the politician's career came when it was revealed that Dastyari warned Huang that his cellphone was likely tapped by American and Australian intelligence services. Hours later, the full audio from the 2016 press conference was leaked to the media, confirming Dastyari's pro-Beijing sentiments.

Meanwhile, Turnbull was set to introduce a legislative package, which would include a ban on foreign political donations. Turnbull went as far as to say that "'Senator Dastyari sold Australia out,' and that the senator was a clear case of someone who took money from people involved in the Chinese government and then "delivered essentially Chinese policy statements." Dastyari resigned a few days later. Huang's citizenship application was denied and he was stripped of his permanent residence while travelling overseas.

Searight also adds that Australia was caught in the middle when China and the United States disputed the origins of COVID-19. She reports China's response to Prime Minister Morrison's suggestion that the World Health Organization (WHO) create a team of independent health inspectors who could be deployed to countries that have outbreaks: "China's Ministry of Foreign Affairs said it was 'gravely concerned' by the 'highly irresponsible' proposal, dismissing it as 'political maneuvering' designed to 'hype up' the situation. Any doubts about China's transparency w(ere) 'disrespectful of the tremendous efforts and sacrifices of the Chinese people.' *Global Times* editor Hu Xijin provided the most colorful example of Beijing's disparaging invective, writing on Weibo: 'Australia is always there, making trouble. It is a bit like chewing gum stuck on the sole of China's shoes. Sometimes you have to find a stone to rub it off.'"

It does not help matters that China's narrative during the pandemic was that Australians were acting as a "lackey" or "pawn" of America. China went as far as to target Australian beef and barley exports, which were valued at AU$2.6 billion and AU$916 million respectively that year. Searight quotes Simon Birmingham, Australia's trade and tourism minister who said, "Australia is no more going to change our policy position on a major public health issue because of economic coercion or threats of coercion than we would change our policy position in matters of national security." She also cites Morrison, who said in mid-May of 2020, "What the Australian government is doing is completely unremarkable. We are standing our ground on our values and the things that we know are always important. ... And those things are not to be traded. Ever."[117] These scandals and the COVID-19

117 Searight, Amy (a senior adviser and director of the Southeast Asia Program

outbreak have caused issues for the Sino-Australian relationship and will have lasting effects for the future.

Despite these recent setbacks, Australia still strives to balance their relations with the US and China. However, Malcolm Cook states that both the US and China have pressured Australia in exchange for strengthening relations. Cook argues that Australia is feeling pressure from both China and America in the South China Sea. China refuses to accept any code of conduct that resembles the arbitral tribunal of 2016 (in which the United Nations Convention on the Law of the Sea declared that China's nine-dash-line claims have no legal basis under international law). The Americans want Australia to conduct FONOPS (Freedom of Navigation Operations) in support for the 2016 ruling. He suggests that Australia should work with other nations in Southeast Asia to negotiate a code of conduct for the region. China's failure to accept the current rules-based order is fueled by their claims to the South China Sea and the way those claims overlap with those of the other powers in the region. Though the US has conducted FONOPS in the South China Sea, the Royal Australian Navy (RAN) has refused to undertake the same exercise. Chinese state media has warned that operations conducted by the RAN would lead to a Chinese attack. Cook elaborates on Australia's view of the UNCLOS ruling by writing, "On the day of the UNCLOS ruling, Australia called on 'the Philippines and China to abide by the ruling, which is final and binding for both parties.'" Despite being encouraged by both the Obama and Trump administrations, Australia refused to do their own FONOPs. In his autobiography, Turnbull said he "considered undertaking such freedom of navigation operations (FONOPs), but was unsure whether the Americans, under Barack Obama, would actually back us if we did."

---

for The Center for Strategic and International Studies [CSIS]). "Countering China's Influence Activities Lessons from Australia." A Report of the CSIS Southeast Asia Program. Published by the CSIS in Washington, D.C. July 2020. 1-3, 5-9, 13, 25-27, 29-30.

Cook argues that Australia should not conduct FONOPs because to do so would require a change in strategy for RAN, and those operations would lead to a reaction from China if Australia were to conduct them rather than America. He recommends that Australia should participate in bilateral multilateral military exercises. Australia already participated in the US-Philippine multi-service Balikatan exercise and the Australia-Vietnam naval exercise during the Australian Defense Force's 2019 Indo-Pacific Endeavor activity.[118] Australia has enjoyed relations with China and America, but both great powers want something from the Australians.

Although Sino-Australian relations were in a downturn from 2017-2021, there was normalizing of relations in 2022. Corey Bell and Elena Collinson argue that after the victory of the Australian Labor Party in the May 2022 election, the rest of the year saw a thaw in Australia's relations with China. They note that Australian Foreign Minister Penny Wong traveled to Beijing that December to participate in the sixth Australia-China Foreign and Strategic Dialogue. Afterward, Wong called the meeting "another step forward as we stabilize the relationship between our two countries." She wrote on Twitter that "Australia believes we can grow our bilateral relationship and uphold our national interests if both countries navigate our differences wisely."

But Bell and Collinson add that Australia still must find ways to balance against China's military, economic and systemic power. While Sino-Australian relations are getting better, Australia is deepening ties with America and Japan to counter China's military. In October of 2022, Australia and Japan signed an updated Joint Declaration on Security Cooperation. Bell and Collinson note that the two nations also committed to "accelerating the consideration of … future rotational deployment of Japan's fighters including F-35s in Australia." China

118 Cook, Dr. Malcolm (Nonresident Fellow at the Lowy Institute from 2010 to 2021. Completed a PhD in International Relations from the Research School of Pacific and Asian Studies at the Australian National University.). "Australia's South China Sea Challenges." Lowy Institute Policy Brief. Published by the Lowy Institute in Sydney, Australia. May 1, 2021. 2-4, 6, 13-14.

has been critical of organizations like the QUAD and AUKUS, both of which Australia are a part of, with China's Foreign Ministry calling the QUAD a "tool for containing and besieging China" and AUKUS an "Anglo-Saxon clique" aimed at building "a NATO replica in the Asia-Pacific."

When Australia acquired a fleet of nuclear-powered submarines, China called it "a violation of the object and purpose of the Nuclear Non-Proliferation Treaty." Australian Defense Strategic Report recommends either procurement or collaboration on hypersonic missiles. In a speech on November 14, 2022, Australian Defense Minister Richard Marles said, "deterrence isn't an alternative to cooperation," and that Australia does not need to "choose between diplomacy and defense" or "cooperation and confrontation." Bell and Collinson write, "In 2014, Australia's then-Prime Minister Tony Abbott rejected US President Barack Obama's request to end Australia's lucrative iron ore trade to China—a proposal labelled 'ridiculous' by Treasurer Joe Hockey. Yet against the backdrop of intensifying geopolitical competition, there is a genuine prospect that Australia could be gearing up to restrict or stymie what is shaping to be a lucrative future export: lithium."

Lithium is a key component in electric vehicle batteries, to the point that Tesla CEO Elon Musk called it the "new oil." China, which dominates the EV-battery industry, purchases nearly 95 percent of Australia's lithium exports and invests in their lithium industry. The authors write that "Resources Minister Madeleine King recently stated that the supply of rare earths 'is as much a national security issue as one of energy and economic security,' and highlighted the need to be 'cognizant of the role Australia's critical minerals will play in the security of our trusted regional friends and allies.' This prompted a strong response from China's Ministry of Foreign Affairs, which said, 'No one should use the economy as a political tool or weapon.'"[119] While

119 Bell, Corey (Dr. Corey Lee Bell is a researcher at the Australia-China Relations Institute, University of Technology Sydney.), Collinson, Elena (head of analysis at the Australia-China Relations Institute, University of Technology Sydney).

Australia is mending ties with China, it will remain a key partner with America.

Another internal problem in Australia is its dependence on China for processing minerals. Natasha Frost states that half of the world's lithium is mined in Australia and China takes most of this. Australia mines 53 percent of the world's lithium and most of it is sold to China. Frost brings up that recently however, the Australian government "wants to break the world's dependence on China for processing the minerals driving the green revolution." China has years of experience and hundreds of lithium refining plants, giving them a head start compared to Australia. Frost cites Marina Zhang, a researcher at the Australia-China Relations Institute at the University of Technology Sydney, who said, "Consumers will vote using their feet, and they will buy electric vehicles, or even solar panels at home, based on the costs."

Pilbara Minerals is working with Calix, an Australian tech company, on a project to refine the mineral spodumene into a lithium phosphate salt. Frost adds that "Dale Henderson, the chief executive of Pilbara Minerals, and other proponents have argued that refining lithium at home would create jobs, reduce the impact of shipping—94 percent of shipped spodumene is thrown out as waste—and secure supply chains for battery chemicals amid rising geopolitical tensions." Refining lithium would allow Australia to get subsidies from America's Inflation Reduction Act, which was the Biden Administration's attempt to cut into China's green energy domination. She writes: "At the Group of 7 summit … (in May of 2023), President Biden and Prime Minister Anthony Albanese of Australia jointly announced projects intended to strengthen the supply chain for 'critical minerals' used in clean energy." The Australian government has invested hundreds of millions of dollars toward the lithium refining industry. Frost cites Allison Britt, a director at Geoscience Australia, a government agency, who said, "If you have more of the supply chain in a country which

"Australia-China Relations: The Outlook for 2023." *The Diplomat*. December 22, 2022. https://thediplomat.com/2022/12/australia-china-relations-the-outlook-for-2023/.

has very strong governance, and a very, very safe and trustworthy business environment, then consumers can have more confidence in the products that they buy."

In 2023, Australia was responsible for 1 percent global lithium refining. A government report estimates that percentage could rise to 20 percent by 2027. Frost cites Albanese, the prime minister, who said in a speech, "I want to make sure that we use the lithium and nickel and other products that we have to make batteries here, that's part of the vision of protecting our national economy going forward." Chris Ellison, the owner of Mineral Resources, said the government must make it easier for foreign companies to invest in Australian lithium refining through incentives like funding and tax breaks which are offered in other countries. "We need the Australian government to come to the party on that," he said.

"Despite the barriers, Australia's control of the raw material gives it a chance to assert influence further down the supply chain," said Joe Lowry, the founder of the advisory firm Global Lithium. "The guy with the rock wins,'" Lowry said. "And Australia's got the rock."[120] Australia's control of its raw resources has caused a split between themselves and the Chinese. That could open doors for other nations to come in and make deals for the raw resources in Australia.

Australia's current issues with China have strengthened their cooperation with the US and other partners in the South China Sea. Jim Gomez and Favila Aaron of the Associated Press covered a May 2024 military exercise in which American and Filipino armies were backed by Australian fighters and surveillance aircraft. Said exercise entailed use of high-precision rockets, artillery and airstrikes, and the sinking of a mock enemy ship. They write that "More than 16,000

---

120 Frost, Natasha (writes *The New York Times*'s weekday newsletter, "The Europe Morning Briefing," and reports on Australia, New Zealand and the Pacific from Melbourne, Australia. Also co-writes the "Australia Letter," a weekly newsletter about the region.). "Australia Tries to Break Its Dependence on China for Lithium Mining." *The New York Times*. May, 23, 2023. https://www.nytimes.com/2023/05/23/business/australia-lithium-refining.html.

military personnel from the United States and the Philippines, along with a few hundred Australian troops and military observers from 14 countries, were participating in annual combat-readiness drills called Balikatan, Tagalog for shoulder-to-shoulder. The drills, … include a scenario of a foreign invasion of the Philippine archipelago. An alliance is very important to show China that you may have all the ships that you have, but we have a lot of firepower to sink all of them."[121]

Australia is also looking to India to help it prevent Chinese encroachment. Associated Press correspondent Rod McGuirk covered meetings in November of 2023, when Indian and Australian defense and foreign ministers met to discuss trade and defense ties. Defense Minister Richard Marles commented after the talks that: "For both of us, China is our biggest trading partner. For both of us, China is our biggest security anxiety."[122] Australia's growing relationship with India provides them a partner that can counter China.

While Australia insists on its military partnership with America, the nation relies on China's economy for their resources, especially their mineral refinery. Australia and the other nations of the QUAD should help each other to become a more coherent force for enforcing peace and stability in the South China Sea. Australia's experience in modern military matters and its cooperation with the American military in the War on Terror provide indispensable knowledge that would help in the scenario of a war. Australia's relationship with China

---

121 Gomez, Jim (Associated Press Chief Correspondent based in Manila, Philippines), Aaron, Favila (Associated Press photographer). "US and Philippine forces sink a ship during largescale drills in the disputed South China Sea." Associated Press. May 9, 2024. https://apnews.com/article/philippines-china-us-military-drills-sea-ship-0157be2b96b90abad85b9c30b29e40a0.

122 McGuirk, Rod (Associated Press staff member based in Canberra, Australia). "India, Australia commit to boosting strategic ties as their diplomats and defense chiefs hold talks." Associated Press. November 21, 2023. https://apnews.com/article/india-australia-talks-ccaecb76ae2d9e029aa80554b55f3e02.

and the United States is one of a careful balancing act between the strongest powers in the world.

Chapter 9: Russia: China's Best Ally or Rivals with a Common Enemy?

Since the Cold War, Russia has been the traditional ally to China. Having the world's largest continuous land border gives the two countries an incentive to work together. Both nations have a vested interest in shrinking America's influence on the world stage, and in this case, it is to drive America out of Southeast Asia. Russia can field naval ships in the Pacific Ocean and participate with Chinese vessels in a time of war. But given that Russia's population is mainly in the part that borders Europe and considering their investment in the war with Ukraine, Russia's options are limited as to how they could help China in a war in the South China Sea.

The storied history of the modern-day Sino-Russian partnership is discussed by Paul J. Bolt. In 1689, the signing of the Treaty of Nerchinsk recognized the Amur basin and the current Russian maritime province as Chinese territory. The treaty also saw licensed border trade between the two empires. Not only was the Treaty of Nerchinsk the first treaty China had with a European power, but it also saw Russian Tsar Peter the Great being recognized by Chinese Emperor Kangxi as a sovereign outside the traditional Chinese tribute system. Bolt cites Harvard historian Odd Arne Westad saying, "Of all the European states, China's first regular foreign relations were with Russia. Indeed, it can be argued that China's first foreign relations—in anything approaching the Western sense—with any country were with that other expanding empire moving into East Asia from the north." Bolt continues with the Qing Dynasty allowing a Russian Orthodox mission into China in 1715. This mission would act as an embassy for Russia and was the only one in Beijing for more than 100 years.

However, in the 1800s, Russia began to take advantage of China's plight at the time. As the Qing Dynasty was dealing with the Taiping Rebellion (1850-1864) and the British and French in the Second Opium War (1856-1860), Russia made territorial demands. In 1858, China agreed to give Russia 185,000 square miles of territory along the northern bank of the Amur. Russia promised to mediate in the Second Opium War, which they did not do. In the 1860 Treaty of Peking, Russia gained 130,000 square miles from China along the coastline. The 1864 Treaty of Tarbagatai saw Russia gaining 350,000 square miles from Xinjiang. The 1860s also saw Russia encouraging Muslim Separatists in Xinjiang. Bolt claims that these moves made China more afraid of Russia than any other European power. He then goes 100 years later, to a time when Mao would call for this land to be given back after all of these "unequal treaties."[123] As these "unequal treaties" are still in place, it is possible that they could become a dividing force between Russia and China, if China either gets what they want from the South China Sea or decides to focus on land claims on the Russian border.

While Russia and China have had relations with each other for hundreds of years, their relationship has been strengthening to counter the Western World. The Sino-Russian relationship in its modern context is discussed by Artyom Lukin. The Sino-Russian cooperation focuses on two important areas: geo-economic and military. This partnership was formed through a combination of Russia's dislike

---

123 Bolt, Paul J. (A professor of Political Science at the United States Air Force Academy in Colorado Springs), Cross, Sharyl N. (Director of the Kozmetsky Center at St. Edward's University in Austin and Global Policy scholar at the Kennan Institute Woodrow Wilson International Center for Scholars in Washington, D.C.), *China, Russia, and Twenty-First Century Global Geopolitics*: China, Russia, and Twenty-First Century Global Geopolitics - Paul J. Bolt, Sharyl N. Cross - Google Books. Published by Oxford University Press. Oxford, Britain. 2018. 3, 7.

of the West's view of them as a defeated adversary and China's humiliation in the Taiwan Strait crisis of 1995 to 1996. While Russia under Putin in the early 2000s tried to strengthen relations with the West, this fell apart in the late 2000s and early 2010s, culminating in the Crimea crisis in 2014 and the accusations of Russian meddling in the 2016 American presidential election. Lukin elaborates that China has two choices: renounce their ambitions or take a stand. If China chooses the latter, then they will have to find strong allies to support their territorial ambitions. Over the course of two decades, the Sino-Russian relationship has grown economically and militarily, according to Lukin: "Since 2009, China has ranked as Russia's top trading partner as an individual country. In 2018, for the first time, Russia–China trade exceeded $100 billion, reaching $107 billion, accounting for 15 percent of Russian exports and imports ... From 2014 to 2019, Russia–China trade volume rose by 17 percent."

In 2015, Russia exceeded even Saudi Arabia as China's top supplier of crude oil, increasing their exports of oil by 60 percent. However, Russia has not attracted the hoped-for economic investment from the Chinese. Chinese investors have a wait-and-see approach to Russia. Much like with other foreign companies, they see a high risk with investing in Russia and find that the reward is not worth said risk. Lukin then goes through the military exercises between the two Asiatic giants. In August 2019, China was invited by Russia to take part in the main strategic maneuvers of that year. Though the exercise's focus was on counterterrorist operations, the Tsentr-2019 drill included repelling airstrikes and counterattacks from a conventionally armed state. In July 2019, Russia and China had a joint air exercise at the South China Sea and Sea of Japan, which involved nuclear-capable bombers accompanied by fighters and surveillance aircraft. The exercise became contentious when South Korea reported that a Russian fighter plane violated Dokdo's (Takeshima's) airspace twice, which led to South Korean interceptor aircraft firing hundreds of warning shots. Lukin then cites Russian Defense Minister Sergei Shoigu saying that the joint air patrol was a symbol to America and its allies in the Pacific that "As two neighbors seeking strategic partnership, Russia and China are thus messaging to everyone that they want to ensure their security." China's Minister of Defense Wei Fenghe had similar sentiments to Shoigu saying, "Let the Americans know about the close ties between the armed forces of China and Russia."[124] Despite a rocky relationship on the economic front, Russia and China are united by a common cause on the military front.

One subject that China and Russia agree upon is their support for North Korea as a buffer between them and South Korea and, indirectly, the United States. Shannon Bugos explains that China and Russia have pushed the UN to lift certain sanctions on North Korea in recognition of denuclearization. However, it is probable that the US would veto such resolution if it were up for a vote. On November 2, 2021, Chinese Foreign Ministry spokesperson Wang Wenbin said at a press briefing that China and Russia aim "to create an enabling atmosphere to facilitate the early start of dialogue" because North Korea "has taken multiple denuclearization measures in recent years, its legitimate and reasonable concerns deserve attention and response." Western powers disagree with that characterization, and contend that North Korea has taken no action to reduce its prohibited nuclear and missile programs. Bugos writes: "During an Oct. 1 meeting, France proposed that the Security Council adopt a joint statement condemning recent North Korean missile tests, but Russia and China opposed the measure."[125] Russia and China enjoy North Korea as a buffer zone between them

---

124 Bolt and Cross.

125 Bugos, Shannon (senior policy analyst at the Arms Control Association, where she conducts research, analysis, and media commentary on U.S.-Russian nuclear arms control, nuclear disarmament and risk reduction, the U.S. defense budget, China's nuclear arsenal, hypersonic weapons, and emerging technologies). "Russia: Russian Officials Talk Nuclear War, U.S. Intelligence Says." :Russia | Arms Control Association. Arms Control Association. Published by Arms Control Association. Washington, D.C., U.S.A. December 2022.

and South Korea, which is more closely aligned with America than with either Russia or China.

While Russia and China are united over their rivalry with the West, they have their own problems to deal with. Sarah Kirchberger, Svenja Sinjen, and Nils Wörmer say that Western observers have looked at the Sino-Russian partnership as no more than a "wary embrace" or a "marriage of convenience." Analysts are concerned about the scope and quality of their cooperation, but it remains unclear how to distinguish between symbolic gestures and actual indications of a deeper strategic trust. "It is important to note here that it would be unrealistic to expect any two countries, and much less countries as large and diverse as Russia and China, to ever arrive at fully convergent priorities, perceptions, and evaluations in all strategically important areas," according to Kirchberger et al. "That is not realistic even with much smaller powers within the EU or NATO, where long-standing animosities and, indeed, tensions can continue to exist, e.g., between Greece and Turkey. Therefore, expectations of China and Russia ever being fully in agreement about their respective national goals, priorities, and perceptions and for them to never experience instances of friction or clashes of interest would be naive."

Both China and Russia are united in their distaste of the Western idea of an international rules-based order. The Kirchberger book cites a study by director of the Carnegie Russia Eurasia Center Alexander Gabuev and a member of Russia's Foreign and Defense Policy Council Dmitri Trenin, who said, "The Russian government conducted a detailed assessment of the benefits, costs, and risks of a closer engagement with China in early 2014 that led to 'the reassessment of two risks: a threat of a Chinese takeover (demographic or military) of Russia's Far East and China's copying of Russian military technology, both of which were determined to be exaggerated or no longer important. In addition, Russian decisionmakers reevaluated Russia's position on China's BRI in Central Asia.'" The same study determined that Chinese and Russian policy and expert communities see the 2014 annexation of Crimea as "a distinct inflection point," and that both countries have a "shared worldview."

The authors then explain that Russia's "Pivot to Asia" was manifested by NATO pushing into Eastern Europe and the imposing of sanctions after Russia's annexation of Crimea. The authors put the recent Sino-Russian relationship in context by comparing it to the manner in which the Taiwan crisis resulted in the United States and Japan agreeing on measures to strengthen their security alliance. In the Yeltsin-Jiang Zemin summit, both Russia and China declared that they would create an "equal, trusting partnership aimed at strategic cooperation in the 21st century." A year later, at the Moscow summit, they would further cement that intention with a "Russian-Chinese Joint Declaration on Multipolarity and the Formation of a New Global Order." The Kirchberger book cites comments by Putin regarding Sino-Russian military cooperation: "We regularly conduct joint military exercises—at sea and on land in China and the Russian Federation—and share best practices in building the armed forces. We have achieved a high level of cooperation in the defense industry. … Undoubtedly, the cooperation between Russia and China increases the defense potential of the Chinese People's Army, which is in the interests of both Russia and China."

The authors then write that Russia has gone to great lengths to decouple from the US Dollar and to connect further with China. "In 2014, when sanctions were imposed, the dollar's share in Russia's export of goods and services ranged between 75 and 80 percent. Subsequently, Russia reached several agreements on trade in national currencies so that by the end of 2020 the share of the dollar in Russia's trade fell to 48 percent. Russia was particularly successful in its trade with China. By 2019, the dollar's share in Russian exports to China had declined to approximately 35 percent and in imports to about 70 percent." Russia's main exports to China are raw materials, fossil fuels, lubricants, and agricultural products. Meanwhile, China's main exports to Russia are electronic products, machinery, vehicles, chemicals, and textiles. However, the authors explain that China has not invested as much into Russia as the Russian government has hoped. "According to (the) central bank (of Russia), the stock of foreign direct investment from China and Hong Kong in January 2019 stood at just US$3.6 billion, or 0.9 percent of Russia's total incoming direct investment ... To conclude, hopes or expectations that the Kremlin may have had for investment winds from China to blow into the sails of the Russian economy and increase its speed have thus far been disappointed. Trade and investments have not only failed to drive the modernization of Russia's economy but served to freeze the authoritarian, centralizing, and kleptocratic features of the economic and political system ... Chinese investors see corruption, red tape, and poor infrastructure as constraints to their operations in Russia."

Chinese officials also debated whether or not to make an alliance with Russia. In 2010, China replaced Germany as Russia's largest single trading partner, and in 2019 Russia became China's 10th largest trading partner. Russia has a smaller GDP than Italy or Canada. They do not have the economic strength to be China's biggest trading partner. The authors summarize that between 2014 and 2015, Russia had 20 Special Economic Zones (SEZs) in the Russian Far East. Since 2019, only six of these SEZs have attracted Chinese investors. Despite the setbacks, Russia is an important partner to China for developing nuclear energy plants, and is one of the largest exporters of crude oil to China. China's defense expenditure in 2020 was four times that of Russia (the same gap shared by America and Britain). Despite that, the book contends that losing Russia as a partner—or worse, having Russia realign with the United States—would be detrimental to China's geopolitical strategy. China and Russia have smaller defense budgets compared to America.

The book claims Russia has not stopped working with China's enemies, as Putin said, "Russia is part of European culture ... I don't imagine my country isolated from Europe ... So, it is difficult for me to see NATO as an enemy." The authors write that "A Russian analyst said 'new challenges and common threats are unlikely to lead in the foreseeable future to the creation of a Russian-Chinese military-political alliance." They cite a Taipei-based Russian specialist, who said, "What Americans have been unable to understand is that Russia, like China, would have gladly traded this Russia-China partnership for better relations with the United States." The authors explain that "The Sino-Russian partnership is fragile, but its potential cannot be underestimated. It can be speculated that in the future, the Sino-Russian partnership may seek to prevent a Western coalition from carrying out an intervention that Moscow and Beijing would deem destabilizing. Because of this partnership, Russia would probably abstain from opposing China if Beijing were to seek a military solution to the Taiwanese stalemate …

"Both Russia and China were elected to the (Human Rights) Council for 2021 until 2023, and the first session of the Human Rights Council in 2021 already indicates close cooperation: out of 30 resolutions passed under the various items in the Human Rights Council session from February/March 2021, 15 were contentious votes and in all cases Russia and China voted in the same manner … In July of 2020 and 2021, Russia and Belarus released a joint statement that dismissed the criticisms of China's policies in Xinjiang. When Russia was under similar scrutiny over their aggression against Ukraine, China alternated from supporting Russia to abstention. … The Chinese-Russian cooperation has gained favor due to other countries being tired of the disproportionate role the West plays in developing countries, the lack of a united West, and the lack of willingness to go against Russia and/or China. However, this cooperation has limits such as the lack of financial support for global initiatives, (and) a lack of constructive proposals for global issues such as the COVID-19 Pandemic."[126]

---

126 Kirchberger, Sarah (the Head of Asia-Pacific Strategy & Security at ISPK. She also serves as Vice President of the German Maritime Institute [DMI] and is a Nonresident Senior Fellow at the Atlantic Council), Sinjen, Svenja (is head of science communication at the Foundation for Science and Democracy [SWuD] and an editor at the journal *SIRIUS*), Wörmer, Nils (is head of the Konrad-Adenauer-Stiftung's

Though Russia and China share common interests, their economic and diplomatic differences have kept them from forming a formal alliance. The Sino-Russian partnership is also in question when it comes to China's response to the Russo-Ukrainian War (2022-). The recent actions of Russia and China are covered by Brian G. Carlson. Carlson contends that Russia is going to be focused on Ukraine for the immediate future, and therefore will have limited reach to influence Asian security affairs. Having been isolated in Europe because of the Ukraine war, Russia has remained diplomatically and economically dependent on China. Carlson explains that even if Russia manages to achieve some of its objectives from the war, its economic and military recovery will depend on China. He goes as far as to claim that the price for said support could lead to Russia going to war in the South China Sea. He adds to his theory by writing that that "Russian military buildups along Ukraine's borders coincided with an increased tempo of incursions by Chinese warplanes into airspace near Taiwan's coastline."

While Russia has supported China's position on Taiwan, China has abstained from punishing Russia and from calling their war in Ukraine an "invasion." Carlson explains that China cannot support Russia too much, lest they face the same international scrutiny as their northern partner. In terms of the Ukraine issue, China remains neutral. Russia's failure to defeat Ukraine quickly could be a cautionary tale for Chinese ambitions in the South China Sea. In 2001, China and Russia signed the Treaty of Good-Neighborliness and Friendly Cooperation, which established the legal basis for their relationship. Since 2013, China and Russia have been doing naval exercises. Carlson writes: "China and Russia have refrained from establishing a formal alliance, partly because both countries have wished to avoid becoming entangled in the other's regional disputes. China declined to endorse

---

International and Security Affairs Department since September 2018). *Russia-China Relations: Emerging Alliance or Eternal Rivals?* Published by Springer. New York City, New York. June 16, 2022. 3, 7, 17-20, 29, 35, 49-51, 54, 109-111, 114-116, 229, 236-238.

Russia's recognition of the sovereignty of two breakaway regions in Georgia in 2008, (or) its annexation of Crimea in 2014 … For its part, Russia officially maintains neutrality on China's territorial disputes in the South China Sea and the East China Sea. The 2001 treaty lacks a mutual defense clause, the crucial feature of an alliance."

China could ask Russia for assistance against Japan and use their airspace to attack Japan. But another thing Carlson notes is that Russian foreign policy in Asia has benefited China's rivals. Russia's most important partner in the South China Sea is Vietnam, which is one of China's biggest rivals. Russia has sold submarines and other advanced weapons to Vietnam. The Russians have also gained access to a naval base at Cam Ranh Bay for servicing and repairs to their warships. If war were to break out in the South China Sea, Russia's access to the naval base would be in jeopardy. Russia had an interest in establishing a base in Vietnam, but that remains unlikely. Because of this relationship, Vietnam has abstained from UN resolutions condemning Russia and its war against Ukraine. Russian officials argued that Vietnam is not anti-Chinese and that their influence is preferable to America's influence. Carlson writes that the Russia-Vietnam relationship has not impeded the deepening of the Sino-Russian relationship. He notes that Russia has enjoyed a warm relationship with India as well.[127] Russia's relationships with China's rivals, along with China's wariness in supporting Russia in the Russo-Ukrainian War, could lead to a divide between the two powers, as their paths diverge from each other.

The lack of a formal Russo-Chinese alliance has given the former breathing room to pick and choose when to lock arms with the latter. Vladimir Portyakov and Alexander Korolev explain that on February 4, 2012, at the Munich Security Conference, Russian Foreign Minister

---

127 Carlson, Brian G. (Former Head of the Global Security Team of the Think Tank at Center for Security Studies [CSS] at ETH Zürich). "China-Russia Relations and Asian Security." :China-Russia Relations and Asian Security - Research Collection (ethz.ch). Strategic Trends 2022: Key Developments in Global Affairs. Published by the Center of Security Studies. Zurich, Switzerland. May 13, 2022. 12-16, 18-19, 29-32.

Sergey Lavrov said, "We will not participate in any arrangements aimed at containing China, which is our good neighbor and strategic partner." After taking office in 2012, Xi chose Russia as his first destination for a foreign trip saying: "Strong China-Russia relations not only answer to our interests but also serve as important, reliable guarantee of an international strategic balance and peace." Portyakov and Korolev explain that after the 2016 ruling of the Hague Tribunal, which found China's claims in the South China Sea unfounded, Russia abstained from taking a side in the matter. As Russian Foreign Ministry Spokesperson Maria Zakharova said on July 16, 2016, "(Russia) had never been a participant in the South China Sea disputes ... would not be involved in them" and "we consider it a matter of principle not to side with any party."

Portyakov and Korolev claim that this shows Russia in a role of a neutral party in terms of the South China Sea, rather than a full-fledged ally: "It took China and Russia almost 10 years before they agreed on prices for delivering Russian gas from Siberia to the neighboring industrial regions of China in 2014. The same is true for the price for electric energy supplies from Russia to China ... These disagreements generate excessive bargaining and an accumulation of dissatisfaction with each other that go beyond energy and can stonewall cooperation in other spheres, as was the case with military-technical cooperation when Russia did not implement an agreement to deliver a large number of military transport aircrafts to China."

The authors write that there is reluctance on both sides of the Sino-Russian partnership in terms of economic cooperation. China views Russia as a source of mineral resources and an export market, which makes the Chinese reluctant to help develop Russia's industrial manufacturing. The fear of Russia becoming a "resource appendage to China" along with suspicion of illegal dumping by Chinese companies, makes the Russian people wary of the evermore increasing power and influence of China. Portyakov and Korolev write: "The two countries' existing economic models generate a relationship pattern that is suboptimal for achieving fruitful cooperation and even dangerous for Russia's eastern territories … The closure of multiple factories and enterprises in Russia's Far East increases unemployment and causes an outflow of an already scarce population." They explain, however, that such fears have dwindled over the years. The FOM (Russia's Public Opinion Foundation) noted that in the early 2000s, more than 40 percent of Russians polled saw China as a threat. In 2014, that dropped to 18 percent. Sergey Karaganov, a presidential advisor to Boris Yeltsin and Vladimir Putin, said that China was viewed as a threat not only by the public, but by the ruling elites too. Karaganov said there was a fear that the Russian Far East would become both an economic and political appendage. This caused Russians to favor containment of China, rather than alignment. In 2001, a report by the Russian Foreign and Defense Policy Council predicted China's economic and demographic rise and called for the prevention of Siberia and the Russian Far East "from becoming a zone of ethnic and socio-political confrontation."[128] While Russia will be partners with China, Moscow remains wary of China's ambitions spilling across the border.While China and Russia have a more-established relationship than many of the other nations aligned with China, the people of Russia, especially the ones living in the Far East, have problems with China. As reported in Gilbert Rozman's Joint US-Korea Academic Studies vol.24|2013, a poll in the early 2010s found that "56 percent of Russian respondents have a sharply negative view of permanent settlement of Chinese migrants in Russia; 35 percent believe that most Chinese visitors end up staying in Russia illegally; and 24 percent think that Beijing has a secret plan to populate the Russian Far East with Chinese migrants."

---

128 Portyakov, Vladimir (is Professor and Deputy Director of the Institute of Far Eastern Affairs, Russian Academy of Sciences, Moscow, Russia). "Reluctant allies: system-unit dynamics and China-Russia relations." :Reluctant allies: system-unit dynamics and China-Russia relations - Alexander Korolev, Vladimir Portyakov, 2019 (sagepub.com). *International Relations* Volume 33, Issue 1, March 2019, Pages 40-66. Published by SAGE Publications. Thousand Oaks, California. March 2019. 43, 51, 53.

Rozman cites a comment by Alexander Abalakov a member of the subcommittee for the development of Siberia: "If we (the Russians) remain passive, we will quickly end up in the position of American Indians on a reservation. We will be fed free of charge and shown to children as representatives of endemic small nationalities, incapable of development."

In 2013, it was estimated by "reliable sources" that 300,000-500,000 Chinese migrants were in Russia, which is far above the 35,000 estimate provided by the Russian Foreign Ministry. Rozman contends that plans to get more investors from East Asia have failed due to the past militarization of the Soviet Union and the criminalization in the Russian Federation. Efforts to solve the labor shortage with migrants have failed due to the unpopularity of Chinese immigration in the Russian Far East.

Though Russia and China want to maintain their partnership, there are avenues Russia can take to protect itself from Chinese influence. Australian national security scholar Alexey D. Muraviev explores avenues that Russia could potentially follow in the event of a breakdown in relations with China. Muraviev claims that, should history repeat itself and relations between Russia and China turn negative, Russia would have to strengthen its defenses on the border with China, deepen its ties with India and Vietnam, and make concessions on any ambitions it may have in Eastern Europe and the Middle East. He mentions that Russian and Chinese ruling elites are well aware of the horror that a war between their two nations would bring, and of the dramatic past that ruling elites from both countries want to avoid.[129] Although both countries want to avoid their collective past, China and

---

129 Muraviev, Alexey D. (Associate Professor of National Security and Strategic Studies at Curtin University, Perth, Western Australia). "Strategic reality check: the current state of Russia–China defence cooperation and the prospects of a deepening 'near alliance.'" :Australian Journal of Defence and Strategic Studies Vol. 3 No.1 (curtin.edu.au). *Australian Journal of Defence and Strategic Studies* Volume 3, Number 1 (2021). Published by the Australian Defense Force. Canberra, Australia. July 1, 2021, 30-32, 35-36, 43.

Russia have problems with sharing Central Asia, and supporting each other when war comes, especially when Russia invaded Ukraine in 2022.

While Russia and China are wary of forming a formal alliance, their relationship does provide a bridge of military cooperation, according to Muraviev. He notes that Russian and Chinese militaries have been collaborating since they signed the Comprehensive defense and military-technological cooperation (MTC) in 1992. In 2016, the amount of Russian military technology within the PLA's inventory was 64 percent, he writes, and "Between 2003 and early 2021, Russia and China staged at least 28 confirmed joint military exercises, 14 of which were bilateral." Muraviev elaborated that since 2009, Russia and China increased their number of exercises, averaging two or three major exercises a year. He explains that Russian open defense sources show that Russo-Chinese naval exercises simulate operations against the fleets of America and its allies. Said sources revealed four scenarios: The Russian Navy engaging the American Navy in either the Atlantic or Pacific maritime theatres; the Russian Pacific Fleet (RUSPAC) engaging with the Japanese Maritime Self-Defense Force (JMSDF) in the Pacific; the PLAN engaging with the USN and JMSDF; and joint battle groups of Russian and Chinese navies operating against the USN, JMSDF and their regional allies."[130] Russia has a military edge on China, which makes the two countries formidable partners when it comes to military exercises and in the scenario of a war with America.

As important as it is to investigate the Sino-Russian relationship, it is also pertinent to delve into Russia's relations with other Southeast Asian nations. Said relations are covered by Eugene Rumer, Richard Sokolsky, and Aleksander Vladicic, who note that Russia has complex political and economic relationships throughout the region. For example, bilateral trade with Japan amounted to only about $20 billion in 2018, while Russia and Japan continue to engage in a contentious territorial dispute over the Kuril Islands. "In 2015, Dmitry Rogozin,

130 Muraviev.

then deputy prime minister in charge of defense issues, responded to Japanese protests against a visit to the Kurils by Prime Minister Dmitry Medvedev by inviting Japanese officials to commit suicide. During his second visit to the islands in 2019, Medvedev casually dismissed Japanese objections and territorial claims as groundless."

The authors also mention that Russia has been fortifying the Kuril Islands with surface-to-ship missiles and airbases. Despite this, Russia has been strengthening its ties with ASEAN nations. Russia and Vietnam have done joint ventures to explore, develop, and produce hydrocarbons. Recent joint ventures like Vietsovpetro, a Vietnamese petroleum enterprise, generate an annual revenue of nearly $2 billion. In 2017, Russian energy company Rosneft and Indonesian energy company PERTAMINA inked a $15 billion deal to build an oil refinery and petrochemical complex in East Java. The authors write that the facility "is expected to become a major regional hub for distribution of Siberian oil throughout Southeast Asia." Russia has also made similar deals for energy projects in Cambodia, Malaysia, Myanmar, Singapore, and Thailand.

Russian arm sales to ASEAN members have increased in the past 15 years as a result of increased military expenditures, restrictions on American arms due to human rights concerns, and a desire to improve ties with Russia to balance against China. However, as a Russian expert on the region said, "Without meaningful economic or trade ties, no mutually beneficial relationship can develop." Rumer, Sokolsky and Vladicic add on to that: "Moreover, Russian inroads in the region notwithstanding, most members of ASEAN continue to value closer commercial and military relations with the United States, primarily because they want to diversify their relationships to maintain a posture of nonalignment, create more room to maneuver among the three great powers in the region, and hedge their bets against an expansionist China."

Despite attempts to deepen relations with ASEAN, as exemplified by the 2016 Russia-ASEAN Summit in Sochi, Russia has been distracted by its ambitions in Ukraine and elsewhere. This has caused the relationship with ASEAN to remain underdeveloped. The authors mention that it does not help that in 2018, Russia's Outward Direct Investment in the European Union was 74 percent, while in ASEAN it was 1 percent.[131] Though Russia has relations with the other Southeast Asian nations, it remains largely preoccupied with spheres of influence in Eastern Europe and Central Asia, rather than in Southeast Asia.

Russia's military capabilities and its reaction to COVID-19 also raise questions about how effective the Sino-Russian partnership would be in the scenario of war. Eugene Rumer and Richard Sokolsky's 2020 paper, "Russia in the Asia-Pacific: Less Than Meets the Eye," showcases some of the cracks in said partnership. Rumer and Sokolsky mention how Russia is lacking in air and naval capabilities, writing: "Although there has been a modest increase in its long-distance force-projection capabilities, Russia lacks the assets and infrastructure to sustain high-intensity conventional operations at long ranges and is no match for the air and naval forces of the United States, especially if combined with the forces of US allies and partners in the region." They also bring up that COVID-19 has exacerbated the anti-Chinese sentiments of Russians living in Siberia and the Far East, to the point that the Kremlin raised the ire of the Chinese government by targeting Chinese nationals in Russia to limit the spread of the virus. This led to Russia closing off the border with China in January of 2020, and China doing the same in April. Rumer and Sokolsky conclude that American leaders, experts, and analysts must assess the possible threat posed by Russia, but must not descend into fear mongering.

131 Rumer, Eugene (senior fellow and the director of Carnegie's Russia and Eurasia Program), Sokolsky, Richard (nonresident senior fellow in Carnegie's Russia and Eurasia Program, worked in the State Department for six different administrations), Vladicic, Aleksander (PhD student in the Department of Government at Cornell University). "Russia in the Asia-Pacific: Less Than Meets the Eye." :SokolskyRumer_Asia-Pacific_FINAL.pdf (carnegieendowment.org). The Return of Global Russia| September 2020. Published by Carnegie Endowment for International Peace. Washington, D.C., U.S.A. 2020. 4, 17, 20-21, 25-26, 30-31, 35.

Russia has been invested in East Asia since the 18th century. However, if Russia joins a war on the side of China, this will have consequences that could cause irreparable damage to the nation. By siding with China in a South China Sea War, Russia would put itself in conflict with India, one of its biggest trading partners. Not only that, but Russia, China, and India are the biggest members of the BRICS economic partnership, which would be split in the event of war. Even in a scenario where Russia wins the war, it would still have to rebuild whatever damage is done to its Far Eastern land, not to mention its land bordering Ukraine. Fighting a war in the South China Sea while fighting in Ukraine would leave the Russian military stretched thin and could cause a collapse on either front. The best-case scenario for Russia is to declare neutrality in the South China Sea War, as this would allow them to reconnect with their economic partners after the war, no matter which side won or lost.

Chapter 10: North Korea: China's Belligerent Neighbor

Since their war with South Korea, North Korea has been a key partner for China in the Korean Peninsula. China has been the lifeline for North Korea ever since North Korea was cut off from the larger world due to pursuing a nuclear weapons program in the 1990s and 2000s. This leaves North Korea with China and Russia as its only major-power partners. Those countries support North Korea for what they perceive as its value in deterring South Korea and indirectly, the United States. Though it is dependent on outside aid, North Korea still has the agency to hold its own missile tests, which scare South Korea, Japan, and America, and contribute to its reputation for being belligerent. This has caused trouble for China, which has been criticized by South Korea for failing to respond after these missile tests. China has been playing both sides as they have economic partnerships with South Korea for economic purposes and with North Korea for military purposes.

To understand North Korea's role in the scenario of a South China Sea War, it is important to investigate their relationship with the two major powers. Stevie L. Jones II covers North Korea's relations with America and China. Jones explains that since the signing of the Korean Armistice in 1953, the United States has failed to have diplomatic relations with North Korea, because they could not find a way around the North's "once an enemy, always an enemy" approach. North Korea also has problems with the US's relationship with South Korea and the memory of American bombing during the Korean War. He goes on to the more recent history between America and North Korea, by elaborating that while North Korea signed the Korean Peninsula Energy Development Organization (KEDO) in 1995, America soon discovered that North Korea had been noncompliant to the bargain. This ended with America suspending oil deliveries to North Korea in the fall of 2002, which gave the North Korean government an excuse to go forth to gain nuclear weapons. With no diplomatic power to pressure North Korea to give up its nuclear ambitions, America has been forced to ask China to intervene. Jones writes: "While China may value the deterrence factor that comes with North Korea's aggression towards the international community, it also recognizes that North Korea's continued aggression will serve only as justification for the United States in pursuing military actions."

China's strategy, according to Jones is: "(to use their) influence to lure smaller power states into establishing economic ties. For China, these small power states represent not only resource security, but they also play a role in the political arena, as they tend to side with China's global policy objectives. For the smaller states, China is a pathway to achieving economic prosperity." Jones cites Dr. Jae Ho Chung and Dr. Myung-hae Choi, who in "Uncertain Allies or Uncomfortable Neighbors" say that North Korea is suspicious of China's intentions, and aware that North Korea's pursuit of nuclear weapons and provocative aggression hampers China's strategic interests. He writes: "North Korea knows China depends on its stability and uses this dependency to be defiant because it knows that China will not cut it off diplomatically or economically."

Jones then cites James Kynge, a principal of China Confidential and Renminbi Compass, who argues that while China is willing to apply some economic pressure on North Korea, the government will not go to the point where the pressure will cause a crisis in North Korea. China cannot allow North Korea to collapse for it would open the floodgates for America. However, other people argue that China has more leverage over North Korea than the other way around. Eleanor Albert, a Ph.D. student in Political Science at the George Washington University said, "China is North Korea's most important ally, biggest trading partner, and main source of food and energy." Sino-North Korean analyst Sabine van Ameijden claims that North Korea is a crucial ally for China, as the North's nuclear tests act as a deterrence to America's presence not only in the Korean Peninsula, but in the Indo-Pacific at large.

Andrei Lankov, a Russian scholar of Asia and a specialist in Korean studies and author of *The Real North Korea*, contends that China sees North Korean aggression as detrimental to stability and long-term goals. A South Korean diplomat told Lankov that "China doesn't have leverage when it comes to dealing with North Korea. What it has is not a lever, but rather a hammer. China can knock North Korea unconscious if it wishes, but it cannot really manipulate its behavior." Jones has a few hypotheses. The first is that China is wary of putting pressure on North Korea because the Chinese government values stability and considers North Korea's activities as a deterrence to foreign influences. Both China and North Korea fear outside influence, as it could lead to instability. The second hypothesis is that China will use its leverage only when North Korea's actions cause destabilization and/or threaten the diplomatic ties between China and other nations. China does not want to be dragged into a war by North Korea. China will use its leverage only when North Korea's actions will lead to war. The third hypothesis is that North Korea will be more likely to defy China in response to outside influences.

North Korea sees outside influence as a threat to their alliance with China, for it could potentially devalue the relationship and drive China to betray them in the future. North Korea can be defiant of China's demands when they perceive it as a threat to the alliance. Jones writes: "For example, North Korea perceives that the United States is attempting to turn China against it by pressuring China to strain its economic relations with North Korea with increased sanctions to press for a non-nuclear North Korea. Either China will be pressured by North Korea's defiance or China will pursue strengthening its relations with the international community."

While China would have loved to characterize the Korean War as a war against imperialism, North Korea's act of striking first against South Korea led to China having to defend North Korea from an America-led United Nations response force.

Jones explains that ever since the 1950s, North Korea has been keen on getting nuclear weapons, with scientists being sent to the Soviet Union for nuclear training in 1956, both countries signing a nuclear cooperation agreement in 1959, and the Yongbyon Nuclear Research center being built in 1964. Despite the cooperation, the Soviet Union refused to give nuclear weapons to North Korea. In 1974, Kim Il-Sung sent a letter to Mao Zedong saying, "As brother countries who shared fighting and dying on the battlefield, China and North Korea should also share the atomic secret." There have been two nuclear crises when it comes to North Korea. The first was in 1992 when America first discovered the nation's pursuit of nuclear weapons, which led to a UN resolution in 1994. In 2002, America discovered North Korea was still pursuing nuclear weapons. The next year, North Korea withdrew from the Nonproliferation Treaty (NPT).

China declined North Korea's nuclear proposal on the ground that investing in a nuclear weapons program in a small country like North Korea would be too costly. Jones cites American academic, author and former foreign policy advisor Victor Cha, saying: "Kim did not want North Korea to be a 'shrimp among whales,' surrounded by a nuclear China, a nuclear Soviet Union, and the United States with nuclear warheads, then deployed in South Korea." Jones suggests that China's greater involvement in the Second Nuclear Crisis was the result of China's aversion to the possibility of getting dragged into another war because of North Korea's aggression. It did not help matters when North Korea was declared by then American President George W. Bush as a nation in the Axis of Evil in his State of the Union Speech in

2002. America needs to work with China to gain more leverage over North Korea.[132] While North Korea enjoys better relations with China than they do with America, the belligerence of the North Korean government has at times been a headache for China as well.

Despite the grievances with each other, North Korea and South Korea can agree on one thing: they are wary of Japan's past and present actions. Said wariness is covered by Patrick McEachern, who explains that Korea's disdain for Japan stems from the years when Japan ruled Korea from 1910 to 1945. During that era, Japan used forced conscription and forced some Korean women into sexual servitude. South Korean leaders recognize that Japan plays an important role in the America-led defense strategy against North Korea. But because of past history and an unwillingness to allow deployment of Japanese troops in South Korea, they refuse to include Japanese officials in talks about military planning. Further complicating the relationship is the reality that both Japan and South Korea claim Takeshima (Dokdo) Island. McEachern contends that both Koreans and Chinese fear that the militarization of Japan will lead Japan to behave in the same manner they did in WWII. China and North Korea have a strategic interest in keeping Japan weak because they recognize that Japan is a critical piece in the American military alliance system.[133] The history Japan and Korea have can complicate relations if war does come to the Korean Peninsula.

While North Korea lacks necessary resources, they are adapting

132 Jones II, Stevie L. (Lieutenant, United States Navy). "China and North Korea: The Leverage Dilemma." :China And North Korea: The Leverage Dilemma (dtic.mil). Naval Postgraduate School: Thesis. Published by the Postgraduate Naval School. Monterey, California. March 1, 2018. 2-3, 13, 15-20, 23, 28-29, 36, 39-41, 50-51, 77.

133 McEachern, Patrick (Deputy Chief, Foreign Policy and Bilateral Affairs Unit, U.S. Embassy Tokyo - U.S. Embassy in Tokyo; Former North Korea Analyst, U.S. State Department's Bureau of Intelligence and Research). *North Korea: What Everyone Needs to Know*. :North Korea: What Everyone Needs to Know - Patrick McEachern - Google Books. Published by Oxford University Press. New York, New York. 2019. 102-104.

their military to modern standards, which could make the dictatorship unwilling to talk of reunification. The military reformation is covered by Mirko Tasic, who argues that "The window of opportunity for successful resolution of the North Korea problem is shrinking, and it seems that not much can be done that could halt further development of North Korea's military capabilities in support of its asymmetric strategy." Artillery and missile systems and nuclear capabilities were the focus of North Korea's military development before 2012, while electronic-warfare and cyber capabilities were the country's focus between 2012 and 2017. He explains that "An asymmetric strategy is one that state and nonstate actors engage in to oppose an adversary of greater military power and capabilities and that targets key vulnerabilities or dependencies of that adversary to create a major psychological impact that affects initiatives, actions, or will."

However, the lack of information and transparency from North Korea makes estimating their military capabilities difficult. Documents prior to the 1990s show North Korea transitioning from conventional warfare to asymmetric warfare, as they pulled out their regular infantry from the demilitarized zone (DMZ) in favor of light infantry. Tasic cites Bruce E. Bechtol Jr., a professor at the Department of Security Studies and Criminal Justice projects at Angelo State University in San Angelo, Texas, who claims "that prior to and during an all-out war NKSOFs might be deployed to carry out asymmetric operations that would include sabotaging lines of communication and taking over command and control centers within South Korea." Bechtol mentions North Korea's development of the Korean People's Navy (KPN), correlates with China's announcement at the 18th Congress of the CCP in 2012 of plans to build up the naval force. He then theorizes that the KPN could be utilized to deliver nuclear weapons, and to use a reflagged commercial vessel to conduct a preemptive attack prior to a war.

Tasic writes that "The transformation of North Korea's military is reinforced by an examination of the growing asymmetric threat posed by capabilities such as short- and long-range artillery, missiles (of short, medium, intermediate, and long range), NKSOFs, and strategic weaponry (weapons of mass destruction, weapons for electromagnetic and electronic warfare, and cyberwarfare capabilities). Andrew Scobell and John M. Sanford's review of North Korea's unconventional forces and capabilities—including nuclear, chemical, and biological weaponry; ballistic missiles; and medium- and short-range missiles and their use in offensive and defensive capacities—suggests that North Korea's focus on strategic weaponry and their asymmetric uses is relatively new. During 2017, a new emphasis on the development of electromagnetic pulse weapons for asymmetric use emerged."

The military incidents involving North Korea in 2002, 2003, 2008 and 2010 could be a way for North Korea to gauge the willingness of American forces. Tasic argues that, given the technological advancements that China and North Korea have made since the Korean War, the United States lacks the capability to conduct a successful military engagement against North Korea. By 2015, North Korea had 1,379,000 total armed forces personnel. This is in comparison to the 838,000 personnel they were sporting in 1985, which made them the world's sixth largest military in terms of personnel. In 2010, North Korea displayed the Hwasong-10 missile, which can attack the first island chain. Throughout 2017, North Korea tested their Hwasong-14 and 15 missiles.

Tasic writes, "A dilemma related to South Korea (ROK) has arisen because of the complexity of North Korea's asymmetric strategy, the true intentions of South Korea's military strategy, and the efforts of the entire international community to understand the North Korea crisis. Consider that two of … ROK's three navy fleets are located on that country's west coast; consider the strength of the ROK air force; and consider that the ROK has a large stock of tanks (2,872), some of which are the most advanced in the world. Then consider that most of North Korea's missile-launch facilities are located on the country's east coast, that the KPN's East Sea Fleet is larger than its West Sea Fleet, and that most KPAAF air bases are in the western part of the country. …

Whether South Korea attacks on the east or west side of North Korea, they run a risk of leaving a flank exposed."

A war with the North is not what the South Koreans want. Since Moon-Jae-in's victory in South Korea's 19th presidential elections "nearly 77 percent of South Koreans believe Seoul should restore dialogue with Pyongyang to help 'resolve' North Korea's nuclear program," Tasic writes. "According to Lieutenant General Wallace 'Chip' Gregson, USA, 'North Korea has adapted to the US-ROK alliance's conventional military superiority by developing tactics and weapons systems that equip them with offensive capabilities that avoid confronting the greatest military strengths of the alliance, in an attempt to compete on what it likely perceives as a more favorable playing field.'"[134] North Korea's military development could prove troublesome for South Korea and America in a South China Sea War.

As North Korea stays dependent on China for resources, it is likely that they would engage on China's side in the scenario of a South China Sea War. Kyu-Seok Shim brings up an example of North Korea's loyalty. When US Secretary of State Mike Pompeo announced that America formally rejected most of Beijing's claims in the South China Sea, and that it would stand with US allies and partners in Southeast Asia to protect their sovereign rights to offshore resources, North Korea responded with the following statement: "Pompeo is outrageously deriding the CPC in such a way as to label the Chinese media as 'propaganda bodies of CPC', COVID-19 as a 'crisis caused by CPC' and 5G telecommunication business as a 'tool of CPC' … 'unbridled criticism' of China was meant to 'sully the trust of the Chinese people in the CPC,' to 'tarnish the international prestige of the CPC' and 'to overpower China with continued harassment from inside and outside …' No matter how maliciously Pompeo slanders the CPC,

---

134 Tasic, Mirko (lecturer and researcher at Webster University's Thailand campus in Bangkok, Thailand). "Exploring North Korea's Asymmetric Military Strategy." *Naval War College Review*, Vol. 72, No. 4 (Autumn 2019). Published by the U.S. Naval War College in Newport, Rhode Island. 53-54, 56-61, 64.

[the criticisms] will only serve to highlight the invincibility of the CPC and the superiority of the socialist system."[135] What this shows is North Korea publicly defending China from American criticism, which could signal their devotion to their alliance with China.

North Korea provides China a bulwark state in the Korean Peninsula in exchange for much needed supplies. They have shown an ability to distract the American alliance with their nuclear saber-rattling. However, there are real questions as to how well North Korea can do in a war against South Korea and the United States soldiers stationed there. It is likely that if North Korea did enter the war, then this would open up another front. China would have to divert materials from the invasion of Taiwan, to prevent American and South Korean soldiers from pouring through their northern border. If China wanted North Korea to enter the South China Sea War, they would send in their own troops and military material to push America out of the Korean Peninsula once and for all.

---

135 Shim, Kyu-Seok. "North Korea blasts Pompeo for South China Sea remarks." *Korea JoongAng Daily*. July 15, 2020. https://koreajoongangdaily.joins.com/2020/07/15/national/northKorea/North-Korea-China-South-China-Sea/20200715174000417.html.

Chapter 11: Pakistan: China's Most Valuable Partner

Because India is China's second-biggest rival, Pakistan has provided China with a much-needed partner to help both militarily and economically in the region of South Asia. Pakistan's antipathy to war with India since their inception make it the perfect partner in the event that China wages a two-front war in the South China Sea and in India. Why is that? Because China risks being blockaded in the South China Sea. Pakistan's position in the Arabian Sea gives China an alternative route to get resources they need.

Among China's friends, Pakistan is the only major power that can open a second front against India, and can move most of its military in the scenario of a large-scale war. Allauddin, Hongsong Liu, and Raja Qaiser Ahmed discuss the history and benefits between these two countries. The authors say that since the establishment of diplomatic relations in 1951, the Sino-Pakistani relationship has grown to a "dependable friendship." They explain that, to China, Pakistan serves as "A link to the strategic maritime outlet to Strait of Hormuz and the Persian Gulf (which) support 20 per cent of the Global Oil trade." Alliance with Pakistan also provides China with "A bulwark against the Indian regional hegemonic ambitions (as well as) achieving peace and stability in Afghanistan which is imperative for greater regional connectivity and also to impede its slipover effects towards fueling militancy and extremism in Chinese Xinjiang province."

China favors a peaceful South Asian region aimed at Pakistan–India economic partnership that would create more substantial economic connectivity. Bilateral trade between Pakistan and China is $9.2 billion.

Allauddin, Liu, and Ahmed calculate that from the mid-'60s to 2011, China invested $5 billion in Pakistan. On April 20-21, 2015, during a trip to Pakistan, Xi signed 51 agreements to "improve trade, commerce and people-to-people contacts, preservation of heritage, communication and private sector investments." In these agreements, Xi pledged that China would give $62 billion in Foreign Direct Investment (FDI) to Pakistan. The bitter relationship between Pakistan and India has been going on since 1948 over the disputed Kashmir region, which led to two wars in 1965 and 1971. This bitterness has led to Pakistan deepening its ties with China.

The authors cite a Pakistani diplomat talking about the 2005 Treaty of Friendship and Cooperation by saying: "The most important aspect of the treaty was the clear and the categorical assurances by China to defend Pakistan's sovereignty, territorial integrity, and political independence." Concerning the problem of terrorism in Pakistan, which remains the most significant threat to economic objectives and prosperity, they write: "China and Pakistan will cooperate regarding terrorism due to three compelling reasons: First, both nations are experiencing separatist movements as China is facing a separatist movement in its western region of Xinjiang which is a Uyghur Muslim majority region that belongs to the militant extremist organization of East Turkistan Islamic Movement (ETIM) and which is responsible for bombing and other violent attacks on Chinese people. … in 2005 both the countries had military exercises regarding terrorism." They also note that China's friendship with Pakistan could push America to deepen its ties with India.[136] Pakistan

---

136 Allauddin (PhD Candidate at School of International Relations and Public Affairs, Shanghai International Studies University in Shanghai, China), Liu, Hongsong (Professor at School of International Relations and Public Affairs, Shanghai International Studies University), Ahmed, Raja Qaiser (Assistant Professor at School of Politics & IR, Quaid I Azam University Islamabad, Pakistan). "The Changing Dynamics and New Developments of China–Pakistan Relations." :The Changing Dynamics and New Developments of China–Pakistan Relations - Allauddin, Hongsong Liu, Raja Qaiser Ahmed, 2020 (sagepub.com). *India Quarterly* Volume 76, Issue 1, March 2020, Pages 73-88. Published by SAGE Publications India. New Delhi, India. March 2020. 73-74, 76-77, 79, 83-86.

and China share both common interests and common enemies, which now bind their relationship into a partnership.

As China and Pakistan's relationship has grown, so too has their economic cooperation. Kiran Hassan claims that the China Pakistan Economic Corridor (CPEC) would be proof of China's win-win foreign policy and of China's commitment to other BRI projects. CPEC was signed in 2017 and is planned to be completed by 2030. The estimated cost of the project is $75 billion, including $45 billion in 2020. Pakistan could act as a different route for China to get oil since the Malacca Strait remains vulnerable for China and 85 percent of their oil imports travel through the strait. The Gwadar port is crucial for China to have better access to the Middle Eastern, African, and European markets. 60 percent of China's energy requirements come from the Persian Gulf. Hassan writes: "The CPEC is often referred to as China's 'flagship' project. China's enduring friendship with Pakistan is one amongst the many reasons why the Chinese feel secure and prioritize the CPEC over many other BRI projects." He estimated that by 2030, the CPEC project will provide 70,000 jobs for Pakistan and 700,000 more in the future. In total, CPEC is projected to create more than 2.3 million jobs, cutting the unemployment rate in Pakistan from 5.5 to 3.9 percent.

Hassan reveals that economic turmoil provides another incentive for Pakistan to accept Chinese economic and infrastructure projects: "Over 140 million Pakistanis either have no access to the power grid or suffer over 12 hours of load shedding daily … Pakistan has suffered from chronic power shortages in the form of load-shedding and power outages in recent years, causing the loss of billions of rupees, severely shrinking the local industry, causing wide-scale unemployment and seriously undermining the Pakistani economy." Hassan adds, however, that Islamic militancy in the western region of Pakistan—especially how terrorist groups are recruiting Uyghurs to their cause—has caused concern for China. "In a visit to China's Shaanxi province in 2001, former Pakistan president Pervez Musharraf met with the imam of the grand mosque in Xi'an and urged all Chinese Muslims to be patriotic, shun violence and work for the good of China." Hassan summarizes the problems with the Sino-Pakistani relationship by writing: "The Chinese will get weary of the many challenges coming with the corrupt, inefficient and globally isolated Pakistan, seriously undermining the conclusion of the Chinese mammoth investment in CPEC. Secondly, Pakistan's prevailing environment of insecurity which is rife with Islamic militancy and domestic insurgency, is thought to be posing serious threats to the construction of the corridor."[137] Though China and Pakistan want to become better partners, that improvement is challenged by terrorism in Pakistan, by China's concerns over potential relations with Pakistani and Uyghur terrorists, and by the political instability within Pakistan. While Pakistan has a good economic partnership with China, they do not share this relationship with their neighbors. Saghir Iqbal writes: "Since independence in 1947 India and Pakistan have gone to war three times, conducted hundreds of other skirmishes and artillery exchanges, and have assisted separatist movements in each other's countries, primarily over the territorial issue of Kashmir." He explains that the problem of Kashmir rose from the indecisiveness of its leader, Maharaja Hari Singh, over whether the state should accede to India or Pakistan. Since gaining independence from the British, India and Pakistan have faced each other in five conflicts (major wars in 1947, 1965, and 1971 and minor actions in 1984 and 1999). Both India and Pakistan are spending more on their military forces since the end of the Cold War. Iqbal elaborates that as recent as 2017, the defense expenditure from both countries was $62.2 billion: $52.5 billion from India and $9.72 billion from Pakistan. Iqbal notes that India has a numbers advantage, with 1.2 million army personnel compared to Pakistan's 560,000.

There is less animosity at the Iranian-Pakistani border, but these

---

137 Hassan, Kiran (University of London · School of Advanced Study), "CPEC: A win-win for China and Pakistan" :CPEC: A win-win for China and Pakistan (degruyter.com). *Human Affairs* Volume 30 Issue 2. Published by De Gruyter. Berlin, Germany. April 15, 2020. 212-218, 221.

two neighboring countries are not partners. Iqbal writes that Iranian and Pakistani relationships have usually gone up and down during the past few centuries. Border tensions due primarily to smugglers and incidence of terrorism have resulted in exchange of fire and mortars, killing security personnel on both sides. For the past 30 years, Pakistan has had increased violence rates, including 1,200 bombings and the Kalashnikov and drug smuggling culture.[138] Pakistan, as formidable as it is, suffers from external and internal problems that can sow disunity in the nation if the pressure of a war comes to their borders.

China's success with investing into Pakistan will be challenged by the country's instability, which has been a problem since the nation's creation. The history of Pakistan disunity is explained by Sangit Ragi, who writes that Pakistan has had trouble maintaining a stable democracy since its inception in 1947. Every brief democratic administration has been replaced with a military regime. Pakistan was envisioned to have a democratic parliament with a federal structure by Quaid-i-Azam (the great leader) Muhammad Ali Jinnah. However, a democratic parliament has collapsed four times in Pakistan's 70-year history.

Ragi writes that "Since 1958, time and again, democracy has been strangled by the periodic and long spells of military rule. The right to freedom, political activities, and civil liberties have been denied and the constitution trampled under military boots." He cites Pakistani diplomat Husain Haqqani, who commented that "Seventy years after its birth, (Pakistan remains) a semi-authoritarian and volatile semi-autonomous, national security state, which has failed to run itself consistently under constitutional order or rule of law." Ragi summarizes Pakistan's issues as follows: "The All-India Muslim League (ML), which

---

138 Iqbal, Saghir (a researcher in International Relations and Security Studies in the University of Lynchburg in Lynchburg, Virginia). *Pakistan's War Machine: An Encyclopedia of its Weapons, Strategy and Military Security* 2018 1st Edition. :Pakistan's War Machine: An Encyclopedia of its Weapons, Strategy and ... - Saghir Iqbal - Google Books. Published by CreateSpace Independent Publishing Platform. Scotts Valley, California, U.S.A. April 13, 2018. 5, 7, 12, 14, 63-64.

took credit for the creation of Pakistan, was unable to transform itself from a nationalist movement into a national party with grass roots organizations that could lead the nation on the path of democracy, constitutionalism, and planned economic development."[139] The lack of a stable government in Pakistan could make it difficult for the nation to wage a war against India.

Pakistan's relationship with China is one that has developed out of their combined animosity for India. Manjari Chatterjee Miller explains the start of Sino-Pakistani relations by writing that "the China-Pakistan relationship was built in the aftermath of the 1962 Sino-Indian border dispute. With little hope of a China-India rapprochement on the border, India became a strategic concern for both countries ... China then provided Pakistan with arms, the materials to build its nuclear weapons program, and large amounts of economic aid. Yet, it never wholeheartedly embraced the relationship ... There were many reasons for this. For one, China never saw India as its equal or a predominant strategic threat. That dubious honor was reserved for the United States. Thus, in the India-Pakistan wars of 1965 and 1971, China did not intervene militarily on Pakistan's behalf.

"Additionally, China became deeply concerned about Pakistan's reputation, somewhat to do with the country's status as a nuclear pariah state after the international outcry about AQ Khan, but mostly for harboring and supporting Islamist terrorists. ... China had long been concerned about Islamist militancy among the Uighurs, and their suspected ties with militants in both Pakistan and Afghanistan exacerbated this concern. ... However, by the mid-2000s, two factors became important. The first was the shift in the US-India relationship. Although the United States had, for a while, been talking about India as a counterbalance to China, it was their progressing military and

---

139 Ragi, Sangit (Sangit Kumar Ragi is currently a professor and Head in the Department of Political Science, University of Delhi.). *Indian Studies Review* Journal Of Centre For Study Of Politics And Governance Delhi Vol. 1 Number 2. July-December 2020 :Publication (cspgindia.com), Published by the University of Delhi. Delhi, India. December 2020. 85-87.

economic partnership in the wake of the collapse of the Soviet Union that hammered this point home for Chinese elites … The second factor was China's ambitions regarding the international order. In 2013, President Xi announced the Belt and Road Initiative (BRI), a gargantuan multilateral infrastructure and investment project, to showcase the Chinese path of development … The relationship raises the specter that India may, in the future, face a two-front war, a scenario that would have been implausible a decade ago."

Miller states however, that the relationship is not problem-free. China's wariness about Islamist militants in Xinjiang and their links to Pakistani militants, its concern about Chinese citizens working in Pakistan who have been the targets of terror attacks, the sporadic opposition in Pakistan to CPEC projects, and China's caution about weighing in on Kashmir all continue to be sticking points. Yet this is no longer just a relationship, but a genuine partnership. India should take note.[140] Despite hiccups in the past and present, China and Pakistan are determined to have a stronger relationship to counter India.

While CPEC has promised economic growth for China and Pakistan, the people of Pakistan have not yet reaped the fruit of this endeavor. The benefits and deficits of CPEC (China- Pakistan Economic Corridor) are discussed by Syed Fazl-E-Haider, who starts by writing that "No doubt the promise was massive. Investment-starved Pakistan has so far reportedly attracted direct investments worth US$25.4 billion from China. Electric power projects under CPEC also fed an estimated 6,000 megawatts into the grid for the energy-deficient nation." Although prospects of CPEC in Pakistan are promising, this project has raised fears over the Pakistani government defaulting on

---

140 Miller, Manjari Chatterjee (senior fellow for India, Pakistan, and South Asia at the Council on Foreign Relations [CFR]. She is also a research associate in the Contemporary South Asian Studies Programme at the Oxford School of Global and Area Studies at the University of Oxford.) "How China and Pakistan Forged Close Ties." *Hindustan Times*. October 3, 2022. https://www.cfr.org/article/how-china-and-pakistan-forged-close-ties.

their external debt with China. Said debt has gone up to $100 billion, with one-third being owed to China.

Though Pakistan has averted the risk of default on its foreign debt by signing a $3 billion deal with the International Monetary Fund, this only came after a tense eight months of negotiations. Pakistan continues to face an acute balance-of-payment crisis and depleting foreign exchange reserves. Fazl-E-Haider adds that locals in Pakistan feel the brunt of this economic turmoil. Inflation increased to 29.4 percent in June of 2023, with food prices up by 40 percent and transport costs by 20 percent over 2022. The poverty rate in the country is estimated to reach 37.2 percent. "And for all the extra energy the CPEC projects have produced, many Pakistanis are struggling to afford the high electricity prices, just as high fuel prices keep people from travelling on the CPEC-built highways." Fazl-E-Haider claims that this is all because a debt-ridden country with a fragile economy is unable to pay long-term development dividends to the masses. He explains that "when Imran Khan came to power in 2018, the brakes went on CPEC. The Khan government wasn't convinced the deal was in Pakistan's national interest and even alleged the then chief minister of Punjab province was taking kickbacks from Chinese companies working on CPEC."

Fazl-E-Haider elaborates that terrorist attacks on Chinese nationals and projects had further stalled CPEC's progress. The strategically located Gwadar Port in Balochistan province is a crucial link but also a security challenge for Chinese nationals. A long-standing local insurgency has morphed to target China's interests with a string of attacks. In 2018, Baloch rebels claimed an assault on the Chinese consulate in Karachi. A year later they took credit for an attack on Chinese tourists at the Pearl Continental Hotel in Gwadar, and in April 2022, a suicide bomber targeted the Confucius Institute at the Karachi University. Also in 2022, a Baloch insurgent group posted a video message on social media warning China: "The Baloch Liberation Army guarantees you that CPEC will fail miserably on Baloch land …

you still have time to quit Balochistan, or you will witness a retaliation from Baloch sons and daughters that you will never forget."

Fazl-E-Haider concluded by writing that "It's not exactly the 'win-win' China or Pakistan might have hoped for—or rivals feared."[141] While CPEC shows a growing bond between the Chinese and Pakistani governments, the people of Pakistan have been split on how the project has affected them and their country.

Along with economic cooperation, China and Pakistan are collaborating militarily as well. Abid Hussain writes that: "China says it will work with Pakistan's military to 'further deepen and expand.'" He then cites an April 2023 statement made by Zhang Youxia, the vice chairman of China's Central Military Commission, to Pakistan's army chief General Syed Asim Munir, on a visit to Beijing. "Zhang said that no matter how the international situation changes, China always gives Pakistan priority in its neighborhood diplomacy." Another statement released by the Pakistan army's media wing, the Inter-Services Public Relations (ISPR), said, "Matters of mutual security interests and military cooperation were discussed. Both military commanders reiterated the need for maintaining peace and stability in the region and enhancing military to military cooperation." The ISPR said Munir would hold further meetings with military officials in China to enhance the "longstanding relations between the two militaries" during his four-day visit.

Muhammad Faisal, an Islamabad-based foreign policy analyst and close observer of Pakistan-China ties, told Al Jazeera that Munir's visit was crucial as it came amid political, economic and security crises in Pakistan. Faisal said, "Of late, Pakistan's dependency on China for economic stability and regional security coordination has grown in the face of financial challenges, renewed threat of terrorism

141 Fazl-E-Haider, Syed (a contributing analyst at the South Asia desk of Wikistrat. He is a freelance columnist and the author of several books, including *The Economic Development of Balochistan*.) "China's big gamble in Pakistan: A 10-year scorecard for CPEC." *The Interpreter*. August 1, 2023. https://www.lowyinstitute.org/the-interpreter/china-s-big-gamble-pakistan-10-year-scorecard-cpec.

and India-centric challenges." November of 2022 saw Prime Minister Shehbaz Sharif making a trip to Beijing and meeting President Xi Jinping. China has invested $60 billion in the ambitious China-Pakistan Economic Corridor (CPEC) project and is Pakistan's key economic and defense partner. Hussain mentions that the South Asian country owes nearly $30 billion—23 percent of its total debt—to China. As Islamabad struggles to resume a much-needed $1.1 billion loan program with the International Monetary Fund (IMF), it has sought help from its allies, mainly China, to roll over some of its existing loans. Analyst Faisal said while the Pakistani military remains engaged with China on regional security, the economy has also taken over as a central agenda in meetings between the military commanders of the two nations.

China continues to help Pakistan economically, even though the last few years have seen multiple attacks on Chinese nationals and facilities carried out by the armed groups in Pakistan. In April 2023, a Chinese national working at a hydropower plant being constructed by a Chinese company in northern Pakistan was accused of blasphemy—a sensitive issue in Muslim-majority Pakistan. The Chinese man was held in judicial custody but released a month later. In 2021, 13 people, including nine Chinese nationals working at the same hydropower project, were killed in an attack claimed by the Pakistan Taliban, known by the acronym TTP. In April 2022, three Chinese teachers and their Pakistani driver were killed in a suicide attack at a university in Pakistan's largest city of Karachi. The attack was claimed by a Baloch rebel group fighting for a separate Balochistan in the southwest. Chinese authorities have repeatedly demanded protection of their nationals amid escalating violence in the country. Hussain comments that "For Beijing, safety and security of its citizens remains the top concern amid resurgence of violence in Pakistan. On this, assurances of army chief will be taken with seriousness." [142] Despite the economic

142 Hussain, Abid (Al Jazeera English's digital correspondent in Islamabad, Pakistan). "China to 'deepen and expand' military ties with Pakistan."

prospects China has in Pakistan, the internal struggles in the country can make it difficult for China to achieve their goals.

Pakistan provides China with a partner that can help them deal with India, their second biggest rival. Pakistan is the only nation in the region that can take on India in a war, which raises the possibility that India could someday face a two-front war with China and Pakistan. The Gwadar port of Pakistan gives China an alternate route for oil and other supplies, which would come in handy in a prolonged war in the South China Sea. Negative issues for China in their relationship with Pakistan include the instability of the Pakistani government, the local economy, and terrorism. Whatever happens to the Pakistani government in the future, it would be wise for them to keep a partner like China around, for economic prospects, and as a deterrent to Indian aggression.

Al Jazeera. April 27, 2023. https://www.aljazeera.com/news/2023/4/27/china-to-deepen-and-expand-military-ties-with-pakistan.

Chapter 12: QUAD: United by a Looming Threat

Given that the neighbors of China have grievances with the major power, it is not surprising that said neighbors would unite to counter the aggression of the Chinese government. The other major powers in the South China Sea region, the United States of America, Japan, India, and Australia formed the Quadrilateral Security Dialogue (QUAD) in 2017. This has resulted in further military cooperation between the four powers to protect free trade in the South China Sea. While they have not explicitly contested China's territorial claims, the Chinese government has complained about the QUAD as an attempt to contain their ambitions. Despite their reluctance to call out China, the QUAD stands as a multi-national organization that could counter China's military.

While QUAD is a new organization compared to NATO and the EU, the seeds for the organization were planted 20 years ago. Sana Hashimi states that QUAD was first established in 2004 to respond to the tsunami that hit India that year. From the four countries, 40,000 troops and responders along with helicopters, cargo planes and transport ships went to give relief to India. Even a Tsunami Core Group was formed to respond to this issue. This response was revolutionary for the time. Marc Grossman, then America's undersecretary of state for political affairs, said that the response to the tsunami was "a new style of diplomacy.

"The Core Group was just such a phenomenon," Grossman said. "This was an ad hoc coalition that ignored traditional groupings. We pulled these specific countries together simply because they were the ones with the resources and the desire to act effectively and quickly." Hashimi then cites then Secretary of State Condoleezza Rice, who had similar sentiments as Grossman, calling it "Transformational Diplomacy." Rice said, "The United States is working with our many partners, particularly our partners who share our values in Europe and in Asia and in other parts of the world to build a true form of global stability, a balance of power that favors freedom."

In 2006, then Foreign Minister to Japan, Taro Aso, said in a speech at the Japan Institute of International Relations: "I firmly believe that Japan must make its ties even firmer with friendly nations that share the common views and interests, namely of course the United States as well as Australia, India, and the member states of the European Union and North Atlantic Treaty Organization (NATO), and at the same time work with these friends towards the expansion of this arc of freedom and prosperity." After a visit to Japan by then Indian Prime Minister Manmohan Singh in December 2006, a joint statement read, "The two leaders [Singh and Abe] share the view on the usefulness of having dialogue among India, Japan and other like-minded countries in the Asia-Pacific region on themes of mutual interest."

The first iteration of the QUAD fell apart after China reacted to the Malabar exercise in September of 2007, calling the military exercise an effort to form an anti-China coalition. After that, Australia remained hesitant on going further with QUAD for the next 10 years. At that time, China sought to address territorial issues with diplomacy and was not as assertive as they have become recently. However, the recent aggression and revisionist policies of the Chinese government caused the nations that make up QUAD to reunite in 2017. The American deputy secretary of state at the time, Stephen E. Biegun, said on August 31, 2020: "Washington's aim is to get the grouping of four countries and others in the region to work together as a bulwark against a potential challenge from China."[143] Through China's assertiveness, the

143 Hashimi, Sana (is Visiting Fellow at Taiwan-Asia Exchange Foundation. She is also an affiliated scholar at Research Institute for Indo-Pacific Affairs, Japan.). "Revival of

four biggest powers in the region managed to form a coalition that can contain China's land claims.

The prominence of the QUAD becomes apparent not only from a geopolitical standpoint, but from a historical standpoint as well. Sangit Ragi explained that in 2017, QUAD went full steam ahead under US President Donald Trump, Australia Prime Minister Malcolm Turnbull, Indian Prime Minister Narendra Modi, and Japanese Prime Minister Shinzo Abe. Since then, members of QUAD visit regularly and work together. Ragi cites former American Secretary of State Mike Pompeo's comment on the goals of the QUAD partnership: "Economic capability and rule of law, protection of intellectual property, trade arrangements, diplomatic engagements, and all the necessary elements which form a security structure, not just military but deeper in nature and extent."

Ragi explains that what made India pursue the QUAD again was China's aggression in the South China Sea. As Indian diplomat Jayant Prasad said: "Beijing's belligerence has been forcing India for a tougher position." President Obama's "Pivot to Asia" and Trump's "Indo-Pacific" partnership signified better integration within their partnerships with the nations of East Asia. Modi's "Act East" Policy emphasized the need for military cohesion with countries near the South China Sea such as Australia. Ragi elaborates that QUAD understands the threat of China, but the members do not have a shared vision of what to do about that threat. From the American Security Project, the researchers write: "Without a clear and unified objective, there runs the risk that the Quad will break up for a second time; in other words, another win for China." Also, none of the members of QUAD are willing to commit to a joint statement on China's aggressiveness in the South China Sea. Trade between QUAD and China makes it harder for the countries to make a unified move against China.[144] While the QUAD can be a

---

the Quad: From Divergences to Convergences.":743f685af69445c89c070adaecd002f1.pdf (mjib.gov.tw). *Prospect & Exploration* Volume 20, Issue 1 84-107. Published by Ministry of Justice Investigation Bureau. New Taipei, Taiwan. 84-87, 89-92.

144 Ragi, Sangit. 37-39, 42-43.

counter to China, the nations in it will have to come up with a decisive stance on Chinese aggression to deter them from going further.

Despite the lack of a joint statement, the members of the QUAD are cooperating to contain China in an event of war. Carolyn Bartholomew and Robin Cleveland cite a comment by Anthony Blinken, who said in Tokyo in March 2021, that "We will push back, if necessary, when China uses coercion and aggression to get its way." In August of the same year, Australian Prime Minister Scott Morrison proposed strategic economic dialogue with America to fight against "economic coercion." In September 2021, the UK, America, and Australia announced the trilateral security pact known as AUKUS, as well as Australia receiving nuclear submarines. The authors elaborate that the Biden Administration built upon the Trump Administration's assertive approach to China over unfair economic deals, threats to national security, and the denial of human rights. Bartholomew and Cleveland write that the US does this by engaging allies and international institutions to enforce the peace and stability in the South China Sea and contain the expansionism of the Chinese government. "In October 2020, a US-India joint statement announced the two countries would begin sharing space situational awareness information. In early 2021, the India Space Research Organization also signed agreements with the Australian Space Agency and Japan Aerospace Exploration Agency to increase cooperation on space-related issues."

The purpose of QUAD has not been lost on China. The spokesperson for China's Foreign Ministry, Zhao Lijian, called the group an "exclusive clique" that "will neither receive a welcome nor have its way." China plans to use Pakistan as a counter to QUAD, especially India. Pakistani Foreign Minister Shah Mahmood Qureshi answered a question on the idea of a PLA naval base in Pakistan with "depends on how the QUAD turns out." Bartholomew and Cleveland discuss how the 2020 border clash between China and India led to Indian policy changes. They write: "In a February 2021 interview, Indian Army Lieutenant General Y K Joshi said India was 'absolutely on the brink' with the

PLA in August 2020." In April of 2021, Indian Ambassador to China Vikram Misri said that Chinese diplomats had a tendency "to sweep (the border conflict) under the carpet and characterize it as just a minor issue," adding that it was "tantamount to running away from the problem."[145] The assertion of China's ambitions has led to the creation of the QUAD and the partnership's efforts to contain the country.

One way QUAD can contain Chinese aggression is through collaboration with other non-QUAD powers in the region. South Korea, Vietnam, and New Zealand make up a group of observing nations (countries that can participate in an organization's events while not being members) called "QUAD Plus." From March 2020 to March 2021, QUAD and QUAD Plus met 15 times to collaborate on delivering vaccines and make efforts to address climate change.[146] ASEAN would also be a great partner for QUAD, as explained by Madeline McLaughlin, the National Security Strategy intern at the American Security Project. She says that this would confront China with the prospect of fighting a war not only with QUAD but also ASEAN. Outside of QUAD, the US can assist ASEAN in building up their own defensive capabilities. The United States can use programs such as the Military Security Initiative and the International Military Education and Training (IMET) to help the South China Sea nations. At the end of July 2020, Vietnam accepted a loan from the US to boost its capabilities of building six new patrol vessels. Building up the defenses of ASEAN will dismantle the Chinese narrative that ASEAN is content with Chinese expansionism in the South China Sea.

---

145 Bartholomew, Carolyn (Chairman of the U.S.-China Economic and Security Review Commission), Cleveland, Robin (Vice Chairman of the U.S.-China Economic and Security Review Commission). "2021 Report to Congress." (The Commission was created on October 30, 2000, by the Floyd D. Spence National Defense). :2021 Annual Report to Congress | U.S.- CHINA | ECONOMIC and SECURITY REVIEW COMMISSION (uscc.gov), U.S.-China Economic and Security Review Commission One Hundred Seventeenth Congress First Session November 2021. Published by U.S.-China Economic and Security Review Commission. Washington, D.C., U.S.A. November 2021. 120.

146 Batholomew and Cleveland. 54, 120, 312-313.

McLaughlin does mention, however, that America should be careful about who they help. For example, Myanmar military has been committing atrocities on the Rohingya minority. This has led to Congress banning Myanmar from the IMET program. When America pulled out of the Trans-Pacific Partnership in 2016, they lost an opportunity to offer nations in the South China Sea an economic alternative to China. To counter this, she states that the United States should seek deeper ties with the nations of ASEAN to prevent them from falling into China's sphere. Chinese State Councilor and Foreign Minister Wang Yi accused America of seeking to disrupt relations between themselves and other countries as a way to isolate China even further. Yi said the Chinese government is on the side of peace and will follow the code of conduct to establish peace in the region.[147] Chinese state media company the *Global Times* warned South Korea that joining QUAD would "inevitably damage the just-restored strategic mutual trust between China and South Korea. ... If Seoul joins the QUAD, it will destroy such mutual trust."[148] Though more nations could join QUAD, China's reaction might result in either military escalation or the degradation of the economy of the nation that joins QUAD.

While QUAD is focused on China's ambitions, the organization is still new and the nations within QUAD will have to invest in further collaboration with each other to keep the organization alive. Abhijnan Rej quotes Indian Naval officer Gurpreet Khurana, who said that the objective of QUAD is as a "Strategic Deterrence vis-à-vis China for individual countries, with the overarching effect on moderating China's behavior, including in terms of adherence to international law and norms of conduct." Rej believes that QUAD should go further and follow the "3Ds": **Dissuade** China from reshaping the

---

147 McLaughlin, Madeline (the National Security Strategy intern at ASP. She graduated from the University of Utah in 2019 with a Fine Arts degree and a political science minor.). "US Strategy in the South China Sea.":resrep26608.pdf (jstor.org). American Security Project (2020). Published by American Security Project in Washington, D.C., U.S.A. October 2020. 4-6.

148 McLaughlin. 320.

Indo-Pacific and from pursuing anti-access/anti-denial policies; **Deter** China from using "grey zone coercion," or to avoid a military response through "ambiguity, asymmetry, and incrementalism.; and **Defend** the freedom and openness of the Indo-Pacific if dissuasion and deterrence fail.

He believes the United States will come to the aid of Japan, Australia, and New Zealand. However, India's involvement in a hypothetical war with China remains uncertain. Although India signed the Logistics Exchange Memorandum of Agreement (LEMOA) with America in August 2016, "no such agreement exists between India and Australia, nor between India and Japan. This is largely due to differences in areas of emphasis for the countries involved. For example, Japan's main interest is in the 'Pacific' part of the Indo-Pacific due to its economic interests while India's is, understandably, in the Indian Ocean region." This is not helped by the questionable quality of India's navy. Rej recounts that during the Malabar exercises between America and India, a former USN officer said, "I assume Indian ships obtained from the prior Soviet Union retain damage control systems/infrastructure that by US standards would be inadequate." Even former Indian Naval officers agreed with this opinion.[149] QUAD can be a significant powerhouse that can counter Chinese aggression, but only as long as the nations can strengthen the military bonds that they have.

The power of QUAD depends on the willingness of its members, especially the United States. Hal Brands states that the United States could follow two models for what a nationalistic grand strategy would look like during and after the Trump presidency. The first model being "Fortress America," a neo-isolationist strategy that would undermine the existing global order in the search for a unilateral advantage. The other model is less destructive of the global structure. In this so-called

---

149 Rej, Abhijnan (a Fellow with Observer Research Foundation's Strategic Studies Programme). "Reclaiming the Indo-Pacific: A Political-Military Strategy for Quad 2.0.":Untitled-1 (orfonline.org). ORF Occasional Paper 147 March 2018. Published by Observer Research Foundation. New Delhi, India. 2018. 8, 10, 14-17.

balance model, America would create better deals that would shift some burden to our partners and allies. Regarding the popularity of "Fortress America," Brands cites a poll conducted in 2013, in which 52 percent of Americans said their country should mind its own business. Brands argues that the balanced economic system can keep America safe and prevent one country from dominating one region by force, since everyone is interconnected. This economic model helps bolster the influence of America and its allies, while rolling back the influence of authoritarian regimes as they get cut off from the international economy. Brands goes on to write, "There may come a time when US primacy has vanished or diminished to the point of strategic insignificance, but that day is still a long way off. And so rather than abandoning a grand strategy that has worked well for the past twenty-five years, America should instead work to sustain and reinvigorate that grand strategy for a period in which its primacy remains impressive, even if more contested than before."[150] While the American Government remains adamant in defense of its interests in the Pacific, the American people will have to be won over to make sure that one of the biggest members of the QUAD is working at its highest capacity in the scenario of war.

While the Russo-Ukrainian War united NATO to the side of Ukraine, the conflict has created a division within the QUAD. Stephen Dziedzic argues that while most members of the QUAD put sanctions on Russia when they invaded Ukraine, India has not done this, nor has India publicly criticized the invasion. In addition, India has abstained from key UN votes to condemn Putin's aggression. Modi has called for the continuation of peace talks in the past, but it is not clear if he has the sway to get Putin to stop his invasion. Dziedzic explains that there are reasons for India's wariness. For one,

---

150 Brands, Hal (American political scientist and writer. He is an international relations scholar of U.S. foreign policy.). *American Grand Strategy in the Age of Trump.* :American Grand Strategy in the Age of Trump - Hal Brands - Google Books. Published by the Bookings Institution Press at Washington, D.C., U.S.A. 2018. 3, 14,16.

the Indian government needed Russia's help with evacuating 15,000 students who were studying in Ukraine. India has had deep military and political ties with Russia ever since the Cold War. As Ian Hall says, "It's not just weapons. There (are) the nuclear power plants Russia is building in India which it wants for its energy needs … It's not unreasonable for them to say, 'we have these security concerns and economic concerns, what are you bringing to us?' … It sounds mercenary, but it's not completely unreasonable … What are we (QUAD nations) willing to do, to fill that gap? Do we trust India enough to give them something like that?"

While the rest of the QUAD was disappointed in India's approach to Russia, America did not expect them to turn on Russia on a dime. Dziedzic then cites the Biden administration's Asia tsar, Kurt Campbell, who said, "We (Americans) understand … India's historic, long-standing relationship with Russia, but at the same time, ultimately, we believe that India will be moving in our direction." Dziedzic summarizes the concerns of other leaders of the QUAD by citing Scott Morrison who said, "We cannot allow what is happening in Ukraine now to ever happen in the Indo-Pacific … We are resolute in our commitment to a free and open Indo-Pacific region where smaller states do not need to live in fear of more powerful ones."[151] Although India is not supporting Russia in their war against Ukraine, this could be a concern as to how they could support their partners in the QUAD and Taiwan in the scenario of war.

As China antagonizes its neighbors over the South China Sea, those neighbors seek a greater connection among themselves to act as a deterrent to China. Geoff Wade notes that Japan is the third

151 Dziedzic, Stephen (is the ABC's Foreign Affairs [Asia Pacific] reporter, based in the Parliament House bureau. He covers foreign policy and Australia's relationship with countries in the Asia Pacific region. Stephen has worked for the ABC since 2007 and spent five years covering federal politics.). "Vladimir Putin's invasion of Ukraine neatly exposes the divisions and limitations of Quad nations." ABC News. March 4, 2022. https://www.abc.net.au/news/2022-03-04/quad-nations-divisions-exposed-by-putins-invasion-of-ukraine/100882258.

largest economy in the world, and the second largest source of foreign investment and third largest trading partner to Australia. It makes sense that these nations would make stronger connections to each other, Wade says: "While the Japan-US alliance is the cornerstone of Japan's national security, the Indo-Pacific region is a key focus for Japan's foreign policy and, since 2017, Japan has been promoting a 'Free and open Indo-Pacific strategy' … Both countries view the Southeast Asian bloc of ASEAN as central to the Indo-Pacific region, and both actively partner with the countries of Southeast Asia in areas including trade, investment, aid, defense, and technology … (Like) Australia, Japan has a strong diplomatic presence in the Pacific and engages at a leaders level through the Pacific Islands Leaders Meeting (PALM)."

Wade showcases the economic connection, as in 2020, trade between Japan and Australia was valued at $66.3 billion. He writes that "From Japan, Australia imports vehicles, refined petroleum, machinery and tires. More recently, Japanese investment into Australia has been expanding into renewables; financial services; infrastructure; information and communications technology; property; food; and agribusiness." He adds that in 2021, Australian investment in Japan was valued at 112 billion Australian dollars or $73.7 billion, American.[152] With Japan and Australia becoming more connected, this could translate into more cooperation between the two of them and the other nations in the QUAD, which would create a stronger deterrence against the Chinese government.

Australia has shown its support for Taiwan and America in the fight for Taiwan's freedom from China. Demetri Sevastopulo writes in the *Financial Times* that "Australia's defense minister has said it was 'inconceivable' that his nation would not support the US in a campaign to defend Taiwan from China, amid rising concerns about Beijing's

---

152 Wade, Geoff (historian who has studied Sino-Asian ties and Chinese influence efforts). "Australia-Japan relations." Parliament of Australia. https://www.aph.gov.au/About_Parliament/Parliamentary_Departments/Parliamentary_Library/pubs/BriefingBook47p/AustraliaJapanRelations.

increasingly assertive military activity … Peter Dutton (an Australian politician) said that Chinese leaders had been 'very clear about their intent to go into Taiwan' and that Canberra had to improve its ability to deter Beijing and be ready to join the US military if it took action." He added, "It would be inconceivable that we wouldn't support the US in an action if the US chose to take that action." Sevastopulo brings these comments into context as they came two months after the launch of AUKUS (a trilateral security partnership among Australia, United Kingdom, and United States) and America's promise to help Australia obtain nuclear submarines. He cites Eric Sayers, a security expert at the American Enterprise Institute, saying that "Australia's rapid strategic realignment on China has been stunning," and "We are now seeing a level of shared tactical clarity emerge in Washington, Canberra and Tokyo on the criticality of stability in the Taiwan Strait." Sevastopulo adds, however, that these comments contrast with the opinion of former Australian Prime Minister Paul Keating, who said that Taiwan was "not a vital Australian interest" and that his country should not be dragged into a military conflict over Taiwan.

In October of 2021, President Biden said that America would defend Taiwan from Chinese attack, which seemed to upend the strategic ambiguity that the United States had on the sovereignty of Taiwan. American policy has generally refrained from saying whether the country would intervene in Taiwan. Sevastopulo says further: "Gerald Brown, an Asia security expert, welcomed Dutton's statement, saying one of the most important factors in deterring Chinese military action against Taiwan was 'creating a multilateral front … Australia's willingness to assist in the event of an unprovoked attack against Taiwan substantially raises the costs for the (People's Republic of China) to engage in hostilities.'" Brown went further to say, "A multilateral front of multiple states willing to step in and support continues to raise these costs and is a prudent move towards deterring PRC aggression." He then cites Anthony Blinken in a call with Japan's foreign minister Yoshimasa Hayashi, in which he called for "working closely with Japan and other allies and partners to advance our shared vision of a free and open Indo-Pacific."[153] Australia can come to the defense of Taiwan, but the divisiveness within their government could hinder their ability to defend Taiwan in the scenario of war.

Such divisiveness is the root of Australia's shifting stance on Taiwan. Rod McGuirk cites Richard Marles, the Australian Defense Minister, who says Australia has made no promises to America that they would fight in a conflict over Taiwan in exchange for American nuclear submarines. In March 2023, President Biden met with Australian and British leaders in San Diego and it was announced that Australia would purchase nuclear-powered attack submarines from the US to modernize its fleet amid growing concern about China's influence in the Indo-Pacific. This was met with criticism by Australians, concerned that America would not provide the five Virginia-class submarines unless Australia promised to be available in the event of a Sino-Taiwanese War. Marles said that Australia had given no assurances over Taiwan. He said, "I want to make it really clear that the moment that there is a flag on the first of those Virginia-class submarines in the early 2030s is the moment that that submarine will be under the complete control of the Australian government." Both Australia and the United States maintain "strategic ambiguity" when it comes to Taiwan. "Marles said while the submarines could be used in the case of a conflict, the main intention was for them to protect vital trade routes through the South China Sea and contribute to regional stability,"[154] according to McGuirk. Australia has been vocal about its desire to take a defensive stance in terms of its position in the South China Sea.

---

153 Sevastopulo, Demetri (Demetri served as Washington Bureau Chief from 2015 to 2021.). "Australia vows to help US defend Taiwan from Chinese attacks." *Financial Times*. November 13, 2021. https://www.ft.com/content/231df882-6667-4145-bc92-d1a54bccf333.

154 McGuirk, Rod (is an Associated Press staff member based in Canberra, Australia). "Australia won't promise to side with US in Taiwan conflict." Associated Press. March 20, 2023. https://apnews.com/article/australia-nuclear-powered-submarines-taiwan-208be03cc40dab2baaeb50843e3c8da7.

However, America believes that Australia would help in a war against China. John Lyons cites American Gen. Mark Milley, former chair of the joint military chiefs of staff, who was asked during an interview with ABC how the American-Australian alliance would be affected if Australia were to refuse to help America in the event of war on Asia. Milley said, "The United States and Australia have been key allies ever since World War I, so well over 100 years and we've been shoulder to shoulder." Lyons explained that senior defense and intelligence figures in Australia hypothesize that in a war with China, Australia could take the port China has at Darwin Port by preventing Chinese ships from entering. He mentions however, that the problem for Australia would be to convince nations like the Solomon Islands to do the same.

India faces a similar problem in Sri Lanka. Lyons cites defense analyst Clinton Fernandes, one of Australia's most-experienced military strategists, who commented: "The fiction is that this is about 'freedom of navigation.' It is not. It is about supporting the US's wish to conduct reconnaissance, intelligence, surveillance, targeting and other military activities in any exclusive economic zone (EEZ) in the world. International law is silent on whether this is legal. Australia and the US say it is. China says it isn't. Many countries that oppose China on other issues happen to agree with it on this issue." He adds that China would target Australian air and seacraft rather than the Australian landmass. Given that 95 percent of trade from China is seaborne, a war would have devastating economic consequences. To reduce tension, Fernandes said, "Come clean with the Australian public. Tell them what we're really doing, and that China does not take a benign view of its encirclement by the US Indo-Pacific Command and Australia. Let parliament vote on whether we should (conduct) these activities. Perhaps the public will agree with the policy—but that has clearly not been tested."[155] While there are doubts about Australia's commitment

---

155 Lyons, John (Global Affairs Editor of the ABC. John has previously been the Editor of *The Sydney Morning Herald*, Executive Producer of the Nine Network's Sunday

to deterring China, they have proven in the past 100 years to be capable of being competent allies with America.

The partners of the Unites States, both in and out of QUAD, have been united by their concern over Chinese aggression. On April 5, 2024, a joint statement by America, Japan, Australia, and the Philippines stated that the four nations are united in upholding the freedom of navigation and overflight, and in respecting the maritime rights given by UNCLOS. To show such commitment, the nations agreed to conduct a Maritime Cooperative Activity (MCA) within the Philippine Exclusive Economic Zone on April 7, 2024. According to the joint statement, "The MCA will be conducted by naval/maritime and air force units in a manner that is consistent with international law as well as domestic laws and rules of respective nations, and with due regard to the safety of navigation and the rights and interests of other States. It will also demonstrate professional interactions among naval/maritime and air forces. Ultimately, the MCA will strengthen the interoperability of our defense/armed forces doctrines, tactics, techniques, and procedures."

Richard Marles, Australia's deputy prime minister and minister for defense, commented: "Australia is committed to working with partners to maintain the global rules-based order. … Australia has consistently emphasized the importance for all states to be able to exercise rights and freedoms, including freedom of navigation, in a manner consistent with international law, particularly the United Nations Convention on the Law of the Sea (UNCLOS). The Maritime Cooperative Activity with the Philippines, Japan and the United States demonstrates our firm commitment to work together to maintain a peaceful, stable and prosperous region."

Minoru Kihara, Japan's minister of Defense, added that "Japan

---

program and Head of Current Affairs and Investigations at the ABC.). "Will Australia support the US in a war between Taiwan and China? Ask the military chiefs, not our political leaders." ABC. August 14, 2022. https://www.abc.net.au/news/2022-08-15/will-australia-join-the-us-in-a-war-between-taiwan-and-china-/101328658.

believes that the issue concerning the South China Sea is directly related to the peace and stability of the region and is a legitimate concern of the international community including Japan, Australia, the Philippines, and the United States, and thus Japan opposes any unilateral changes to the status quo by force, such attempts as well as any actions that increase tensions in the South China Sea."

Gilberto C. Teodoro Jr., secretary of national defense of the Philippines, added to the statement that "The Comprehensive Archipelagic Defense Concept (CADC) (is what) we are implementing includes (the) strengthening and deepening cooperation and interoperability with all nations, big and small, to maintain regional peace and stability as well as good order at sea based on international law, principally UNCLOS … This first in a series of activities demonstrates the enduring friendship and partnership among the peace-loving peoples of the Philippines, United States, Australia, and Japan."

American Secretary of Defense Lloyd J. Austin III concluded the joint statement by saying, "Every country should be free to conduct lawful air and maritime operations. These activities with our allies Australia, Japan, and the Philippines underscore our shared commitment to ensuring that all countries are free to fly, sail, and operate wherever international law allows. Our operations together support peace and stability at the heart of our shared vision for a free and open region."[156] This unity over Chinese aggression will help the QUAD gain more partners for the near future.

Not only does the QUAD have the ability to counter China's military, but it also has the chance to grow in the South China Sea. Given the grievances the nations of the South China Sea have with China's overlapping territorial claims, the countries within and without the

156 Austin III, Lloyd J. "JOINT STATEMENT: Australia – Japan – Philippines – United States Maritime Cooperative Activity." April 4, 2024. https://www.defense.gov/News/Releases/Release/Article/3733095/joint-statement-australia-japan-philippines-united-states-maritime-cooperative/.

QUAD have a unifying issue that the four nations of the organization can use to their advantage. The joint military exercises demonstrate that the QUAD will have more cohesion when working together, in comparison with China, which only has a history of exercising with Russia. If a war were to occur, the QUAD could use its cooperation and cohesion to get the upper hand in their fight against China.

Chapter 13: ASEAN: A Thorn in China's Side

Although China's and the United States's disputes over Taiwan and the Korean Peninsula get the most attention, the two major powers also vie for influence over the Association of Southeast Asian Nations (ASEAN).

Made up of Indonesia, Vietnam, Malaysia, Brunei, Laos, Cambodia, Thailand, Myanmar, Singapore, and the Philippines, ASEAN is an organization that has promoted neutrality since the Cold War. The association also has been responsible for mediating negotiations for territorial claims between its member nations. China's claim over all of the South China Sea overlaps with many members of ASEAN. Not only that, but China has a history of using its fishing and Coast Guard vessels to assault ASEAN vessels. Meanwhile, most ASEAN nations have important and long-standing economic relationships with China. That makes it hard for ASEAN to decide on a response to Chinese expansionism.

While ASEAN nations work with each other to negotiate overlapping claims in the South China Sea, China's actions in the region have made mediation difficult. Richard Q. Turcsányi covers the claims and actions of the nations of ASEAN. Turcsányi states that Vietnam, the Philippines and Malaysia have claims over the Spratly Islands. Leaders of Malaysia and Vietnam visited their parts of the Spratlys in 2010 and 2011 respectively. Turcsányi writes that China comes into the story in 2011 and 2012, with actions that included cutting cables on seismic survey vessels of Vietnam. China also deployed surveillance vessels to protect its fishermen who were fishing in the maritime territory of the Philippines. China rejected claims of any wrongdoing on its part and asserted that Vietnam and the Philippines escalated the dispute by operating in the area belonging to China.

Turcsányi also covers how China used economic measures to achieve its political goals. China decided to cut its banana imports from the Philippines, officially citing quality issues. This resulted in a significant decrease of Philippine bananas in the Chinese market, both in terms of quantity and market share. On 2 May 2014, China deployed the Haiyang Shiyou 981 oil rig in the disputed waters near the Paracel Islands, on the Vietnamese side of the median line between the Chinese and Vietnamese coasts in what China claims as its EEZ generated by the Paracel Islands. Thirty Vietnamese boats tried to intervene, but they were prevented from doing so by more than 80 vessels of the China Coast Guard. One of the direct outcomes of the incident was massive anti-China protests in Vietnam, which led to damages to Chinese property and deaths of Chinese nationals. Similarly, this event contributed to Vietnamese strategic considerations and further deepening of Vietnamese relations with the USA and other actors, possibly balancing China's power.

Since 2014, China engaged in massive reclamation projects and the construction of artificial islands in at least seven locations in the Spratly Islands and at least three locations in the Paracel Islands. The construction included seawalls, airfields, multi-floor buildings (some as tall as 10 floors), radar towers, gun emplacements, ship docks, and helicopter bases. In June 2015 China announced that the land reclamation in the South China Sea would end soon. China is not the only country that engages in reclamation and construction works in the SCS—Vietnam and Taiwan recently conducted such activities as well. However, the Chinese construction is unparalleled in its scope. According to the Pentagon, by August 2015 China reclaimed 2900 acres of land, Vietnam 80 acres, Malaysia 70 acres, the Philippines 14 acres, and Taiwan 8 acres. Newly constructed outposts enable deployment of Chinese Navy vessels. At least three locations (Mischief Reef, Subi

Reef, and Fiery Cross Reef) have received PLA Navy frigates.[157] Though every nation in the region has a claim on a piece of the South China Sea, China decided to make moves against all other claimants rather than to divide and conquer.

Though ASEAN has had a rocky relationship with China in recent years, individual nations that make up the organization have their own feelings about China and other regional powers. Premesha Saha discusses these differences. Laos and Cambodia have closer relations to China, while Singapore, Vietnam, and Myanmar are more invested in India as a security provider. Meanwhile, the Philippines, Thailand and Brunei remain neutral on this security issue. Saha explains that, given the tension in the region, ASEAN centrality and unity is waning. However, she believes India could provide help by participating in military exercises with ASEAN. Saha writes: "India has already been engaged in capacity-building initiatives with countries like Vietnam and the Philippines. India and Vietnam pursue naval cooperation which includes composite training programmes in the field of submarines, aviation, and dockyard training. In 2015, both countries' coast guards signed an MOU for establishment of collaborative relationship to combat transnational crime and for sharing best management practices and augmenting operational-level interactions."

These exercises help not only in the cooperation between India and ASEAN, but also in enhancing India's ability to deter Chinese bullying in the South China Sea. She contends that the Indian Coast Guard can give training to the coast guards of Vietnam, Indonesia, and the Philippines. The Indian Coast Guard can provide assistance in dealing with illegal fishing, a thorn in all Southeast Asian countries' sides.

Saha then discusses how Indonesia seeks American investment

---

157 Turcsányi, Richard Q. (an Associate Research Fellow at the Institute of International Relations Prague who specializes in Chinese foreign policy, China's power and China's relations with Central Europe). *Chinese Assertiveness in the South China Sea: Power Sources, Domestic Politics, and Reactive Foreign Policy*: 978-3-319-67648-7.pdf (springer.com). Published by Springer. New York, New York, U.S.A. 2018. 44-51, 53.

in the development of the Riau Islands's Natuna regency, which is in Indonesia's outer islands in the South China Sea. She suggests that India and the rest of the QUAD nations can help mediate the territorial claims that China and ASEAN have on the South China Sea. The significance of the ASEAN group is highlighted by the fact that in the first quarter of 2020, ASEAN passed the EU as China's largest trading partner.[158] While ASEAN remains neutral, the nations within the organization are choosing to stick to the status quo, or to collaborate with either China or the QUAD.

As mentioned before, the South China Sea remains messy with several nations claiming different parts of a single island chain. Examples such as the Spratly Islands and the Scarborough Shoal are covered by Mark E. Rosen, who writes that "The Philippines has asserted claims to Scarborough Shoal as well as a collection of 50 (Spratly Island) features which are known collectively as the Kalayaan island group (KIG). The Philippine claims to sovereignty over the features known as Scarborough Shoal and the KIG are independent of its archipelagic status both legally and historically." He cites Michael McDevitt, then the American acting deputy assistant secretary of defense for East Asia, who said, "The reality on the ground is that China, Taiwan, Vietnam, Malaysia, and the Philippines all permanently occupy features in the Spratly group; some have done so for over 50 years."

Rosen states that the Scarborough Shoal has become a point of contention for China and the Philippines, since it is rich in fish, guano, sea turtles, sea cucumbers, and other living marine resources. China takes the position that the Scarborough Shoal (Huangyan Island) is a part of the Zhongsha Islands and could be entitled to a full 200-nautical mile EEZ. Rosen brings up a response to an op-ed in *The Wall Street Journal*, by a Chinese Ministry of Foreign Affairs official

---

158 Saha, Premesha (a Fellow with ORF's Strategic Studies Programme). "India Calibrates its South China Sea Approach.":ORF_IssueBrief_477_India-SouthChinaSea_FinalForUpload.pdf (orfonline.org). Issue Brief No. 477 July 2021. Published by Observer Research Foundation. New Delhi, India. 2021. 14-17.

who said, "Hundreds of years of jurisdiction has consolidated China's sovereignty over the island. Historic and legal evidence are explicit, clear, complete and thorough, as proved by official documents, local chronicles and official maps throughout Chinese history. China's sovereignty over Huangyan Island has long been recognized and respected by the international community and had not been disputed by the Philippines. On the contrary, the Philippine claim over the island has never been recognized by any other country. The claim that the Huangyan Island belongs to the Philippines because it's closer to the country is a result of ignorance of the international law."

Rosen notes that Philippine claims to sovereignty and jurisdiction over the rocks of Bajo de Masinloc are not premised on proximity or the fact that the rocks are within its 200-nm EEZ or CS (Continental Shelf) under the UN Convention on the Law of the Sea (UNCLOS). Although the Philippines exercise sovereign rights over its EEZ and CS, nonetheless, the reason why the rock features of Bajo de Masinloc are Philippine territories is anchored on other principles of public international law … In the case of Bajo de Masinloc, the Philippines have exercised both effective occupation and effective jurisdiction over Bajo de Masinloc since its independence … Bajo de Masinloc was also used as an impact range by Philippine and US Naval Forces stationed in Subic Bay in Zambales.

Though most people will side with either the Philippines or China over the Spratlys, Rosen reveals that some, including Raul "Pete" Pedrozo, an international legal expert, contend Vietnam should own the Spratlys. Pedrozo explained his position in *The Issue of South China Sea* in June 2000. He said China's historical claims are based on sporadic visits from Chinese fishermen and a collection of archaic charts. The actions of the fishermen cannot be considered actions by a sovereign government unless said actions are followed by official actions. Speaking of which, the Chinese naval patrols are insufficient for China to claim authority on the feature. China never made formal protest over France's assertion of sovereignty in the Spratlys in 1933.

Only in 1935 did the Spratlys appear on Chinese maps, which only contested the claims of the French. China did not protest France ceding the Spratlys to Vietnam in 1950-1951. In terms of legality, the moves on the Spratlys by Taiwan, China, and the Philippines are not only considered legally ineffective, but also forcible occupation. Rosen mentions that it does not help the Philippines that America takes no stance on their claims on the Spratlys or Scarborough Shoal.[159] The Spratlys and the Scarborough Shoal are just a couple examples to showcase the confusing nature of the South China Sea claims.

The nations of ASEAN have also partaken in the modernization of their militaries to defend themselves and their claims. Dereck Grossman claims that Vietnam's military modernization is focused on deterring China from taking disputed territory in the South China Sea. As a part of this modernization, Vietnam has acquired Kilo-class submarines, Su-30MK2 maritime strike fighters, and a network of anti-access missiles. But if deterrence fails, Grossman argued that Vietnam would not be able to take on China in a drawn-out fight: "This conclusion is supported by an analysis of its evolving warfighting doctrine in the air and sea domains, lack of jointness and unrealistic training activities, and shortcomings in the Vietnamese military's maritime domain awareness capabilities and weapon systems interoperability."

Grossman elaborates that in November 2017, before the Asia-Pacific Economic Cooperation (APEC) meeting in Da Nang, Vietnam, China's Xi Jinping issued a statement on how well Vietnam and China worked together and how they had a history of "good neighborliness." However, Vietnamese officials have been concerned that China's actions may destabilize the status quo. The South China Sea is Vietnam's economic lifeline. The country has not only interest in coastal development and resources in their EEZ, but the sea is also where a million Vietnamese

---

159 Rosen, Mark E. (Mark Rosen is an SVP and General Counsel at CNA Corporation). "Philippine Claims in the South China Sea: A Legal Analysis": Philippine Claims in the South China Sea: A Legal Analysis (cna.org). A CNA Occasional Paper. Published by CNA analysis and solutions. Arlington County, Virginia, U.S.A. August 2014. I-II, VII, 7-12, 20-22.

fishermen and their 120,000 boats operate. Grossman explains that territorial disputes over the South China Sea are nothing new. In 1974, China took the Paracel Islands from Vietnam. In 1988, they then took the Spratly Islands from Vietnam. Grossman cites Vietnamese Senior Lieutenant General and Deputy Defense Minister Nguyen Chi Vinh, who said at the Shangri-La Dialogue of June 2016, that Vietnam seeks to cooperate with China "to find common points in strategic interests." Vinh went on to say that China is responsible for "changes to the status quo along with the threat of militarization."

While Vietnam knows that it cannot match China's technological or numerical advantage in the sea or the air, Vietnam seeks a more modest route of modernization to deter China from seizing any disputed land in the South China Sea. Grossman writes that Vietnam has increased its coast guard presence and now has a larger coast guard than the Philippines, Malaysia, and Indonesia combined. While Vietnam has probably achieved its purpose of posing enough force to deter China from attack, Vietnam had never fought a war like this before. Grossman explains that one proposal by the VPA (Vietnamese People's Army) is the "all-people's war," or "the mobilization of the entire population to protect the nation." This would give the VPA the ability to conduct asymmetrical warfare. Vietnamese leaders anticipate the use of guerrilla tactics in what has been called the "people's war in the new situation."

However, Grossman brings up how the Soviet Union (now Russia) and China supported North Vietnam against the Americans in the Vietnam War. That support no longer exists, though Vietnam is working with Russia, India, and Japan in the realms of technology transfers and coproduction. The VPA is also working on making a stockpile of weapons to reduce the need for supply from other countries. Another problem that Vietnam faces is the exalted position the army has over the navy and air force. Though the position of the Vietnamese army is earned through their victories over the United States in 1975 and China in 1979, this neglects the VPN (Vietnamese People's Navy) and

VAD-AF (Vietnam Air Defense-Air Force). Both organizations have been derided as "the army's ships and the army's planes." Despite the greater emphasis on training throughout recent years, there has not been a noticeable improvement in the sea and air forces. The VPN has limited proficiency and confidence in close-in maneuvers, and all exercises are being held at long distances. For VAD-AF, according to Nguyen Thanh Trung, a fighter pilot and national war hero: "for a military pilot, accumulating 1,000 flight hours is very difficult and time-consuming, maybe (taking) 10 to 15 years, or even much longer." This means that a Vietnamese pilot only receives 65-100 flight hours a year. The lack of flight hours is due to previous accidents and the lack of expertise in high-tech weapons.

However, Grossman highlights that, in late 2016, Vietnam accepted India's offer to train their Su-30 pilots. Japan has also deepened its partnership with Vietnam. In January 2017, Japan sent six used coast guard boats to Vietnam. In June of that same year, both countries had their first joint exercise to fight against illegal fishing in the South China Sea. Grossman notes a significant rekindling of relations between the United States and Vietnam. After Obama's visit to Vietnam in May 2016, America lifted the embargo on arm sales, which resulted in America encouraging the VPA to consider buying MDA assets.

"In November 2017, President Donald Trump stopped for two days in Vietnam, spending one day in Da Nang to address APEC and the second to meet with top Vietnamese leaders in Hanoi. Trump's visit reassured Vietnam of the US position in the South China Sea, following up on Prime Minister Nguyen Xuan Phuc's visit to the White House in May 2017. The meeting was significant because it reiterated the need for Washington to cooperate with Hanoi on issues of 'maritime security'—code for primarily dealing with China in the South China Sea." Grossman recommends that the United States should seek greater defense and military cooperation with Vietnam so that Vietnam would have better chances in a military confrontation with

China.[160] Vietnam's ability to take on China will have to be bolstered by its partners if there is to be any chance of pushing China back in a scenario of war.

Given that Vietnam is one of the few ASEAN nations with a land border with China, there is a more nuanced predicament between China, Vietnam, and the other powers in the region. Said predicament is covered by Le Hong Hiep, who writes: "Since independence, Vietnam has pursued a two-pronged strategy to handle a preponderant China: on the one hand, Vietnam has shown its unwavering determination to thwart Chinese attempts to undermine its political autonomy or territorial integrity. On the other hand, Vietnam has also paid due deference to China as long as its own independence and autonomy were respected." China has been Vietnam's biggest trading partner since 2004. In 2011, trade between the two countries was valued at $35.7 billion, more than a thousand times larger than trade between the two in 1991. Hiep argues that the interdependence China has with Vietnam has prevented them from sanctioning Vietnam or eliciting concessions. However, Hiep also writes: "By September 2013, Vietnam had established strategic partnerships with Russia (2001), Japan (2006), India (2007), China (2008), South Korea, Spain (2009), the United Kingdom (2010), Germany (2011), Italy, Thailand, Indonesia, Singapore, and France (2013)." He notes that Vietnam has entered into "comprehensive partnerships" with Australia in 2009 and America in 2013. Hiep argues that the closer relationship with America has given Vietnam more confidence to take on China, given their mutual interests to curb Chinese aggression.[161] Vietnam's land border with

---

160 Grossman, Dereck (a senior defense analyst at the RAND Corporation)."Can Vietnam's Military Stand Up to China in the South China Sea?": Can Vietnam's Military Stand Up to China in the South China Sea? on JSTOR. *Asia Policy*, Vol. 13, No. 1 (JANUARY 2018), pp. 113-134. Published by the NBR. Seattle, Washington, U.S.A. January 2018.

161 Hiep, Le Hong (Dr. Le Hong Hiep is a Senior Fellow at the Vietnam Studies Programme.). "Vietnam's Hedging Strategy against China since Normalization." : Vietnam's Hedging Strategy against China since Normalization on JSTOR. *Contemporary*

China and its location on the South China Sea give Vietnam the ability to have beneficial relations with both America and China. Vietnam would prove to be a valuable ally in the event of war between the two global powers.

The cautious nature of ASEAN's response to China's intrusions into their EEZs (Exclusive Economic Zones) does not address the present and future needs of the people of ASEAN. Such needs are discussed by Daria S. Panarina, who mentions that in 2015, oil demand in the South China Sea reached to 4.4 million barrels a day (mb/d). It is projected by the Economic Research Institute for ASEAN and East Asia that demand will increase by 4.1 percent per year to reach 12.2 mb/d in 2040. The biggest demand will come from Indonesia as it is estimated to increase by 4.8 percent per year, to 4.8 mb/d by 2040. In 2015, ASEAN produced 2.9 mb/d of crude oil. However, as Panarina writes: "according to The Institute of Energy Economics, Japan (IEEJ) estimates, (ASEAN production) is likely to decrease to 2.1 mb/d in 2040. The region's import dependency was 44 percent in 2015, which is already significant, and it could rise to 83 percent in 2040."

The Energy Information Administration (EIA) has estimated that there are 11 billion barrels (bbl) of oil and 190 trillion cubic feet (Tcf) of natural gas reserves in the South China Sea. However, Panarina also cites energy consultancy Wood Mackenzie, which contends that there could be only 2.5 bbl of oil in reserves. To uncover the murkiness, Panarina writes: "The US Geological Survey (USGS) analyzed the potential for undiscovered conventional oil and gas fields within several geologic provinces of Southeast Asia in 2010 as part of its World Petroleum Resources Assessment Project. The study included a significant area of the South China Sea, which as the USGS estimates may contain anywhere between 5 and 22 billion barrels of oil and between 70 and 290 trillion cubic feet of gas in as-yet undiscovered resources (not including the Gulf of Thailand and other areas adjacent

---

Southeast Asia, Vol. 35, No. 3 (December 2013), pp. 333-368. Published by ISEAS–Yusof Ishak Institute. Singapore. December 2013. 334, 345-346, 357, 359.

to the South China Sea). These additional resources are not considered commercial reserves at this time because it is unclear how economically feasible it would be to extract them."

The countries in ASEAN will likely have to look to other alternatives. As Panarina writes: "The Philippines generates about a third of the electricity for its main island of Luzon from a single source, the Malampaya gas field, which is expected to cease production by 2027. Unless an alternative is found—and Reed Bank is the only good option currently on the table—the Philippines will need to either import significant amounts of natural gas at greater costs, rapidly incorporate other energy sources into its power supply, or face severe shortages."

She moves from oil to fish, as 55 percent of global marine fishing vessels operate in this region, with 12 percent of catches taking place there. The fishery faces significant challenges, as Panarina explains: "Total fish stocks in the South China Sea have been depleted by 70-95 percent since the 1950s and catch rates have declined by 66-75 percent over the last 20 years. Giant clam harvesting, dredging, and artificial island building in recent years severely damaged or destroyed over 160 square kilometers, or about 40,000 acres, of coral reefs, which have already declined by 16 percent per decade."[162] The need for the energy and food for the people of the South China Sea is becoming a time bomb, as the negotiation for said resources is lacking.

ASEAN's neutrality comes both from recent events and the history of the organization's founding. Ralf Emmers explains the beginning of ASEAN neutrality by writing: "The Association of Southeast Asian Nations (ASEAN), established in Bangkok in August 1967, has sought to peacefully manage the territorial disputes involving four of its ten members, as well as China and Taiwan, while seeking to preserve

---

162 Panarina, Daria S. (Researcher, Center for Southeast Asia, Australia and Oceania, Institute of Oriental Studies, Moscow, Russia). "Why South China Sea is Important for the Philippines: Economic Aspect.": SEA2020n1(46)p356-369.pdf (ivran.ru). *Southeast Asia: Actual Problems Of Development* 2020, Volume I, No. 1 (46). pp. 356-369. Published By Institute Of Oriental Studies Russian Academy Of Sciences. Moscow, Russia. January 2020. 357-358, 360, 362-363.

its diplomatic neutrality on the sovereignty issue." Countries within ASEAN would rather talk directly to each other than negotiate through the auspices of the association. In 2002, ASEAN signed the Declaration on The Conduct of Parties in The South China Sea, which commits the organization to behave: "without resorting to the threat or use of force, through friendly consultations and negotiations by sovereign states directly concerned, in accordance with universally recognized principles of international law."

Emmers explains that the Scarborough standoff had immediate effects on ASEAN. Those effects were seen in the ASEAN Ministerial Meeting (AMM) held in Phnom Penh in July 2012, where discussion on China took place. Being dependent on economic development and aid from China, ASEAN member Cambodia called for appeasing China and minimizing the internationalization of the South China Sea. This resulted in an unusual disagreement. Elmers writes that, for the first time in 45 years, "ASEAN failed to issue a joint communiqué due to differences over the South China Sea question. The Philippines had asked for the Scarborough Shoal incident to be mentioned in the draft communiqué, but Cambodia refused to include a reference on the grounds that the sovereignty disputes with China are bilateral and not multilateral in nature."[163] Although ASEAN neutrality is an important cornerstone of the organization, the encroachment by China will make keeping neutrality difficult.

Although China and ASEAN have butted heads in the past, the interdependence they have with each other could deter either from waging war. Said interdependence is explained by Wai Ting, who claims that the economic integration between China and ASEAN leaves any land claims in a paradox. He explains that this integration started in the early 1990s and has been growing ever since, from nearly $8 billion

163 Emmers, Ralf (Senior Lecturer in International Politics of East Asia and Co-Chair of the Centre for International Studies and Diplomacy [CISD] in London, Britain). "ASEAN's Search for Neutrality in the South China Sea.":04_Ralf Emmers_DOI.indd (snu.ac.kr). *Asian Journal of Peacebuilding* Vol. 2 No. 1 (2014): 61-77. Published by IPUS at Seoul National University. Seoul, South Korea. 2014. 62, 64, 67.

in trade in 1991, to nearly $300 billion in 2010. He supports that point by writing: "Approximately 2,700 ships pass through the waters (of the South China Sea) every day. 50 percent of Asia's oil supply, including 60 percent of Japan's oil import, and 80 percent of Asia's strategic goods also pass through the SCS."

Ting delves further into the importance of the South China Sea, by mentioning that the fishing output in the South China Sea is 8 percent of the world's total output and is made up of 7 million tons of fish with an estimated value of $6.5 billion. Each year, 70,000 ships go into the South China Sea, carrying one-third of the world's trade. Ting explains that ASEAN may have to choose between China and the United States soon, and the stakes for a conflict in the South China Sea would be severe.[164] Despite their weariness with the situation, ASEAN nations are aware that their geographical location makes them vulnerable to the actions of the world's great powers.

As a rising power in the region, Indonesia is as weary of Chinese encroachment as nations in both ASEAN and QUAD. Indonesia's actions against Chinese encroachment in the early-to-mid 2010s are covered by Arianto A. Patunru, Mari Pangetsu, and M. Chatib Barsi. They explain that even while Indonesia in 2015 sought foreign investment from China, President Joko Widodo stood firm in his opposition to Chinese actions in the South China Sea, which led to contention over the Natuna Islands. After winning the presidential election in July 2014, Widodo talked about Indonesia being an intermediary force in the South China Sea. He later told a Japanese newspaper that Indonesia rejects China's nine-dash line as having no legal basis in international law.

Sino-Indonesian relations were further soured by three incidents

---

164 Ting, Wai (Emeritus Professor at Hong Kong Baptist University, Hong Kong). "Increasing Economic Interdependence between China and ASEAN and Its Implications." :Conflicts-in-the-South-China-Sea-and-China-ASEAN-Economic-Interdependence-A-Challenge-to-Cooperation.pdf (researchgate.net). *Journal of ASEAN Studies*, Vol. 4, No. 2 (2016), pp. 127-141. Published by CBDS Bina Nusantara University. Kebon Jeruk, Indonesia. 2016. 127-128, 132-133, 139.

between Chinese fishermen and the Indonesian Coast Guard from March to June of 2016. In June that year, the Chinese foreign ministry spokesperson said that China has "no territorial sovereignty issue" with Indonesia, but that the confrontation dispute was about "Chinese fishermen's traditional fishing grounds and where China and Indonesia have overlapping claims for maritime rights and interests." This was China's first use of the term "overlapping claims" when it comes to Natuna.[165] The actions of Indonesia against Chinese aggression could prove to be a unifying factor between QUAD and the biggest nation within ASEAN.

However, as explained before, not every nation in ASEAN has negative views on China. Malaysia is one country that holds sympathetic views to China as covered by Peter Kreuzer, who argues that, unlike the Philippines, Malaysia has few issues with China when it comes to territorial claims. He backs this up with historical context, as Malaysia was the first of the ASEAN nations to open diplomatic ties to China in the 1970s. The Prime Minister at the time, Tun Abdul Razak, said in 1971 that China had "the right to play a role in regional affairs commensurate with her importance and dignity as a major power and the right to expect that the countries of the region do not act in ways which adversely affect her." Since that time, Chinese and Malaysian leaders have held high-level meetings with each other in their respective countries. An opinion piece published by both *The People's Daily* in China and Malaysia's own *New Straits Times* stated that: "to Razak: 'Malaysia and China share a similar vision. The 'Chinese dream' is of economic prosperity, peace and happiness; a

165 Patunru, Arianto A. (A fellow at the Arndt-Corden Department of Economics and Policy Engagement Coordinator), Pangetsu, Mari (Professor of International Economics at the Faculty of Economics in the University of Indonesia in Jakarta, Indonesia), Basri, M. Chatib (Senior Lecturer at the Department of Economics in the University of Indonesia in Jakarta, Indonesia). *Indonesia in the New World: Globalisation, Nationalism and Sovereignty.* :Indonesia in the New World: Globalisation, Nationalism and Sovereignty - Google Books. Published by IEAS Yusof Ishak Institute. Singapore. 2018. 4, 58.

dream of opportunity not instability." Chief of the Malaysian Armed Forces, General Zulkifeli said that "non-claimants (America) must take caution on their actions especially in the disputed areas which may cause unwanted provocations and hinder the progress of the negotiations."[166] Nations like Malaysia prevent any further moves of support between ASEAN and QUAD.

Despite how close it is to China, Vietnam has expressed disdain for the Chinese government and its aggression towards Vietnamese claims. Le Hong Hiep's journal article discusses the recent Vietnamese grievances against Chinese encroachment. Ever since normalizing relations with China in 1991, Vietnam had a cordial relationship with their northern neighbor up until recently. In 2004, China became Vietnam's biggest trading partner, and in 2008 both countries announced the foundation of a "comprehensive strategic cooperative partnership." Vietnam wants to maintain peace with China, but the Chinese government is making that increasingly difficult. After the 2014 oil-rig crisis, Vietnam began to perceive China as a threat to their security. At the end of 2014, Vietnam had defense and strategic dialogue with America, Britain, France, Italy, Russia, India, China, South Korea, Singapore, and Australia. Vietnam's defense ties with America, Japan and India make China both wary and irritated.

This military policy is not lost on China, as the 2004 Chinese Defense White Paper reads: "Vietnam's policy is not to join any military alliance, not to allow any foreign country to establish military base in Vietnam, and not to take part in any military action that uses force or threatens to use force against another country … Vietnam is not going to undertake arms races but constantly strengthens its

166 Kreuzer, Peter (Senior Research Fellow at the Peace Research Institute Frankfurt in Germany). "A Comparison of Malaysian and Philippine Responses to China in the South China Sea." :Comparison of Malaysian and Philippine Responses to China in the South China Sea | The Chinese Journal of International Politics | Oxford Academic (oup.com). *The Chinese Journal of International Politics*, Volume 9, Issue 3, Autumn 2016, Pages 239–276. Published by Oxford University Press. Oxford, Britain. June 16, 2016. 239, 249-250, 265, 269-270.

military capabilities for the purpose of sufficient self-defense." Hiep argues that the danger of joining an alliance in the Cold War is what cements Vietnam's unwillingness to join an official alliance. That, and they do not want to anger China.

China's aggression has made it harder for Vietnam to keep its "no alliance" policy, and the American side has stepped in to support Vietnam's interests. In July of 2014, the US Senate passed a resolution (S. RES.412), which called on China to withdraw the oil rig and accompanying ships, and Deputy Assistant Secretary of State Michael Fuchs called for a "freeze" in China's provocative actions. Meanwhile, apart from statements dismissive of China's actions during the crisis, Japan also announced in August 2014 that it would provide Vietnam with six patrol vessels as part of an aid package.[167] Vietnam is reluctant to join an alliance now, but China's push into territory claimed by Vietnam and support from the QUAD nations, could move Vietnam to choosing a side in a war in the South China Sea.

China's actions against Vietnam in the South China Sea not only angered the government, but also the people. The crisis and the damage that it caused is explored by Phuong Hoang. After a Chinese surveillance vessel cut the cables of a Vietnamese ship conducting seismic research in the South China Sea in 2011, hundreds of Vietnamese protested against China. In the oil-rig crisis in 2014, anti-China protests only got larger and more violent. The Vietnamese government did not take an aggressive stance on China and suppressed the demonstrations before they could do too much damage to Sino-Vietnamese relations. The crisis led to 21 people dying and another 100 suffering injuries. Hoang notes that anti-Chinese protests in Vietnam turned violent in 2012 and 2013. Ten thousand people protested in Vietnam, including more than 2,000 in front of the Chinese embassy in Hanoi and 5,000

---

167 Hiep, Le Hong (Dr. Le Hong Hiep is a Senior Fellow at the Vietnam Studies Programme and the Regional Strategic and Political Studies Programme of the ISEAS – Yusof Ishak Institute.). "Vietnam's Alliance Politics in the South China Sea.":trends_in_sea_2015_6.pdf (iseas.edu.sg). *Trends in Southeast Asia* 2015 #06. Published by IEAS. Singapore. 2015. 2-5, 9.

in Ho Chi Minh City. Seven thousand workers demonstrated in front of Chinese businesses and 20,000 workers rioted against Chinese-owned factories and mistakenly attacked Taiwanese and South Korean factories. After two days of rioting, 15 factories were destroyed, and 460 other factories were looted and vandalized. Despite the violent protests, this did not change Vietnam's approach to China. However, since the early 2010's Japan, India, and America have deepened their ties to Vietnam, with the American carrier the USS *Carl Vinson* visiting Vietnam in 2018 for the first time since the Vietnam War. Both Japan and India have become key security partners for Vietnam, which has received equipment and training from the two countries.[168] Although the Vietnamese government wants a good relationship with China, this has become more difficult because of the actions of the Chinese government, and the anger of the Vietnamese people.

While China has rocky relations with most of the ASEAN nations, countries like Cambodia have been more favorable toward China. Monineath El explains that Cambodia has good relationships with both China and Japan. This goes as far as China and Japan enjoying a strategic partnership with Cambodia since 2010 and 2013, respectively. Though Cambodia supports China in the South China Sea, they remain neutral in what El calls "the East China Sea War."

"Cambodia, as a developing country, does not want to take sides in any regional and international disputes, for example East China Sea dispute, and would prefer to stay neutral," he writes.

Chinese investment in economic development and infrastructure-building comes at the cost of Cambodian gold, timber, and land concession

---

168 Hoang, Phuong (National Intelligence University, Washington, D.C.). "Domestic Protests and Foreign Policy: An Examination of Anti-China Protests in Vietnam and Vietnamese Policy Towards China Regarding the South China Sea." :Domestic Protests and Foreign Policy: An Examination of Anti-China Protests in Vietnam and Vietnamese Policy Towards China Regarding the South China Sea - Phuong Hoang, 2019 (sagepub.com). *Journal of Asian Security and International Affairs* Volume 6, Issue 1. Published by SAGE publications. Thousand Oaks, California, U.S.A. March 26, 2019. 2, 7, 9, 13, 15, 18, 20.

at a low restriction on development. Since 1994, Japan has been the largest donor of aid for economic development, infrastructure, national building, and peace building in Cambodia. In May 2012, an agreement on military cooperation was signed by then Chinese Defense Minister Liang Guanglie and present Cambodian Defense Minister Tea Banh, which allowed Cambodian cadets to train in China. China also sent $17 million worth of military aid to Cambodia and built a training facility in the country. In January 2013, China donated military vehicles and set up a training program in auto-repair. El notes that in another agreement signed in 2013: "China would enhance the capacity and expertise of the Royal Cambodian Armed Forces by offering training courses and providing military equipment and materials."

Japan approved a $2-million package for their military engineers to train Cambodian troops. This was an attempt by Japan to build up military relationships with neighboring countries as a way of defending their claims on the Senkaku Islands. Chinese people make the second biggest group of tourists in Cambodia and Chinese investors control 90 percent of the garment manufacturing industry in Cambodia. Chinese investment in Cambodia accounted for 3,016 businesses and an investment value of $1.58 billion. In comparison, Japan has $328 million of investment in Cambodia.[169] This interdependence with both China and Japan could lead to Cambodia facing economic danger if Sino-Japanese relations worsen.

While ASEAN has stayed neutral in the rising tensions between America and China, individual nations within the organization have been wearied by China's intrusions into their territories. From Chinese fishing boats overfishing in other countries, to the Chinese Coast Guard harassing other naval vessels, the nations of ASEAN have a list

---

169 El, Monineath (Tax Economist at Ministry of Economy and Finance at Phnom Penh, Cambodia). "Cambodia-between-China-and-Japan-in-east-China-sea-dispute." :Microsoft Office 2000 (researchgate.net). Royal University of Phnom Penh Department of International Studies Foreign Policy II: Cambodia Focus. Published by Royal University of Phnom Penh in Cambodia. 2015. 3-5, 9-10, 15-17, 19-20.

of grievances with China's aggressive tactics. Given that ASEAN has made weapons deals with America and even had officers trained by American officers, most of the nations in ASEAN have an established military relationship with the United States. However, with Laos and Cambodia having deeper relations with China than America, it is possible that ASEAN will split in two if war comes to the region.

Chapter 14:
Europe: Willing, but Are They Ready?

Ever since the late 1400s, Europe has wanted to establish trade in the Indo-Pacific to get spices and other valuable resources. In the modern day, Europe is still invested in the region and is interested in the politics of the Indo-Pacific. More specifically, many European nations are worried about the growing tension between America and China in the South China Sea. The European Union (EU) and Britain are investing more into the defense of their holdings in the South China Sea and have strengthened their ties with countries in the region to help counter China. Despite the long distance between Europe and the South China Sea, Europe is keen on keeping their interests in the region safe.

Much like America, Europe's relationship with China had soured in the past decade. Stephen M. Walt, in a 2021 *Foreign Policy* article, claimed that Chinese ambitions in the South China Sea have caused a shift in European attitudes. He backs this up by writing that "Twenty years ago, European attitudes toward China were mostly positive; today, the share of the population with an 'unfavorable' view of China has reached 63 percent in Spain, 85 percent in Sweden, 70 percent in France, and 71 percent in Germany. China's confrontational diplomatic style and its efforts to impose cultural uniformity by 'reeducating' millions of Uyghurs have played poorly in Europe as well."

This makes European governments wary of working with China. China's aggression will lead to Europe and America, especially America's allies in Europe, to unite under the ideas of trade and human rights. However, Europe has only formed coalitions against other European powers, such as Napoleonic France or Germany in both World Wars. Walt cites French Finance Minister Bruno Le Maire, who said, "The United States wants to confront China. The European Union wants to engage China." Europe also wants to be "independent from the United States, able to defend its own interests, whether economic or strategic interests." Some European nations could help maintain balance against China. France, for example, has a presence in the South China Sea with more than a million citizens in the Indo-Pacific. Walt concludes by writing that "Europe might join a US-led balancing coalition because it understood that Americans won't go on protecting them if they opt for neutrality vis-à-vis China."[170] Europe's reaction to China's aggression and to atrocities against the Uyghur people might push them to take a more proactive approach in the South China Sea.

While the EU is limited by geography to affect what goes on in the South China Sea, the organization has continued to strengthen connections with nations in the region that China has bullied. The EU has made strides in forging a relationship with Japan, which is explored by Dimitri Vanoverbeke, Takao Suami, Takako Ueta, and Frederik Ponjaert. Since its formation in 1998, the Japan-EU Conference has played an important role in connecting two of the strongest powers in the world. The modern relationship between Japan and the EU is separated into 2 waves. The first wave (1990-2005) was based on economics; the second wave (2006-) is based on military defense.

In 2013, Japan and NATO launched a "Joint Political Declaration between Japan and the North Atlantic Treaty Organization." This included plans to cooperate on cyber defense, disaster relief, counter terrorism, disarmament and non-proliferation, and maritime security. In July 2013, Japan signed with Britian the Agreement Concerning the Transfer of Arms and Military Technologies Necessary to Implement

---

170 Walt, Stephen M. (a columnist at Foreign Policy and the Robert and Renée Belfer professor of international relations at Harvard University). "Will Europe Ever Really Confront China?" *Foreign Policy*. October 15, 2021. https://foreignpolicy.com/2021/10/15/will-europe-ever-really-confront-china/.

Joint Research, Development and Production of Defense Equipment and other Related Items. France signed on to the agreement in March 2015. With this connection, the EU has become more involved in the politics of the South China Sea. The EU has spoken out against violence and the use of force. Following a series of violent anti-Japanese demonstrations in China, in 2012, EU High Representative Catherine Ashton said, "The EU urges all parties concerned to seek peaceful and cooperative solutions in accordance with international law, in particular the UN Convention on the Law of the Sea and to clarify the basis for their claims. The EU calls on all parties to take steps to calm the situation." In 2015, a joint declaration by the G7 Foreign Ministers (which included European and Japanese representatives) said: "We continue to observe the situation in the East and South China Seas and are concerned by any unilateral actions, such as large-scale land reclamation, which change the status quo and increase tensions. We strongly oppose any attempt to assert territorial or maritime claims through the use of intimidation, coercion, or force."[171] The military and economic connection that Japan has with the EU, could provide support for Japan in the event of war.

Given that they are some of the biggest economies in the world, the EU and China have growing economic ties with each other, which makes the EU's relationship with the nations of the South China Sea complex. Joshi Manoj emphasizes the economic interconnectedness

---

171 Vanoverbeke, Dimitri (Has a PhD from KU Leuven and lectures in Japanese Studies at the Department of Area Studies at Leuven), Suami, Takao (Professor of Law at Waseda University Law School and served as the President of the Japanese Association of EU Studies), Ueta, Takako (professor at the International Christian University in Japan and was Ambassador, Deputy Chief of the Mission of Japan to the EU), Peeters, Nicholas (Studied Law at the University of Innsbruck and was the Minister/Deputy Head of the EU Delegation to Japan from 2002 to 2006), Ponjaert, Frederik (Researcher and Lecturer at the Institute of European Studies at the ULB and KU Leuven). *Developing EU–Japan Relations in a Changing Regional Context: A Focus on Security, Law, and Policies.* :Developing EU–Japan Relations in a Changing Regional Context: A Focus on ... - Google Books. The Globalisation, Europe and Multilateralism Series. Published by Routledge, New York, New York. 2018. XII, 2, 14-16, 22-23.

between the EU and China: "In 2018 Chinese exports to the EU were valued at Euro 394.7 billion, while those to the US, Euro 406.317 billion; meanwhile, EU exports to the US were Euro 406.372 billion in the same year, and to China, Euro 210 billion." A European Commission document published in March 2019 mentions that China is "a cooperation partner with whom the EU has closely aligned objectives, a negotiating partner with who the EU needs to find a balance of interests, an economic competitor in pursuit of technological leadership and a systemic rival promoting alternative models of governance."

In early 2018, Bloomberg audited a decade of Chinese purchases, analyzing data for 678 completed or pending deals in 30 countries since 2008. The audit found that both Chinese state-backed and Chinese private companies had been involved in deals worth at least US$255 billion across the continent, including the takeover of some 360 companies. This has caused countries like France and Sweden to change their laws on company takeovers.[172] Though the EU wants economic ties with China, they do not want China to control the economies of the nations of the EU.

The EU's connections with nations in the QUAD and the South China Sea in general can be the foundation for a support-based system for the nations that will be on the front lines of a war against China. However, the internal problems in Europe could impair the EU's ability to support the QUAD and other nations in the South China Sea. Liselotte Odgaard states that internal issues such as Brexit, growing populism, and the migration crisis have already tested the strength of the EU. European Council President Donald Tusk said in May 2018 that "the EU had to be 'prepared to act alone' without the United States." Odgaard brings up that: "The independent policy line was reflected

---

172 Manoj, Joshi (a Distinguished Fellow at the Observer Research Foundation, New Delhi). "China and Europe: Trade, Technology and Competition.":OP#194-new text (orfonline.org). ORF Occasional Paper #194 May 2019. Published by Observer Research Foundation, New Delhi, India. May 2019. 3, 6, 9, 11, 16, 19-20.

in the April 2019 EU-China summit statement. The agreement was a fait accompli readily accepted by Beijing, which was keen on a joint statement due to its conflicts on trade and market economic regulation with the United States." This is in contrast to next year as "In 2019, France agreed with Japan and Australia to share common supplies such as food, fuel, and ammunition, an agreement that Japan already had with the United States and Australia and is currently negotiating with India."

Singapore sought a partnership with the EU partly to help deal with economic and security issues that result from the presence of America and China. To Singapore, the EU is a gateway to further cooperation with ASEAN. France went as far as participating in bilateral exercises with the United States, Japan, and Australia. Odgaard explains, however, that France, Italy, and Germany are reluctant to adopt America's confrontational stance on China.[173] The European Union's independent streak may keep them out of the war with China; considering their strong ties with the QUAD and Southeast Asian nations, they might have to make a choice.

Despite leaving the EU in 2016, Britain agrees with other European nations that the future lies in Southeast Asia. Britain's military investment into Southeast Asia is covered in a 2018 book by Houlden Gorden, Scott N. Romaniuk, and Nong Hong. With the fifth largest economy in the world and the sixth largest military budget, Britain is an important player on the world stage. When Britain voted for Brexit on June 23, 2016, the country decided to champion the "Global Britain" initiative. More specifically, "'Global Britain' is about reinvesting in our relationships, championing the rules-based international order and demonstrating that the UK is open, outward looking, and confident on the world stage," according to the authors.

---

173 Odgaard, Liselotte (A Senior Fellow at the Hudson Institute at Washington, D.C.). "European Engagement in the Indo-Pacific." :European Engagement in the Indo-Pacific on JSTOR. *Asia Policy* Vol. 14, No. 4 (OCTOBER 2019), pp. 129-160. Published by National Bureau of Asian Research, New Delhi, India. October 2019. 135-137, 153-154.

Britain has had the "Pivot to Asia" policy since 2010, when then Prime Minister David Cameron and Chancellor George Osborne pursued a closer relationship with China. But the British government under Theresa May invested in relations with Vietnam, South Korea, and Japan. A conflict in the South China Sea would disrupt trade going to and from Britain, which explains why the UK has been focusing on Asia for the past decade. There is a debate over the effectiveness of the British military presence in the region. Some have argued that any long-term benefit may be outweighed by the short-term risk of angering China.

However, Britain's interest in the South China Sea is not a recent issue. As far back as 1907, senior British diplomat Sir Eyre Crowe wrote a memorandum on this subject, suggesting that Britain should be deployed to protect the smaller nations in the area and protect their independence. British historian John Bew and British political scientist David Martin Jones repeated Crowe's sentiments: "The first priority of UK strategy in Asia must be to ensure that these alliances are maintained and, where possible, bolstered—particularly in South and Southeast Asia. A strategic stance tuned to twenty-first century realities, rather than appeals to historical sentiment, will provide a far more effective basis for utilizing Commonwealth links to India, Pakistan, Malaysia, Singapore, Australia, and New Zealand."

Going into the modern day, with the United States being responsible for these protective duties, the British see themselves having to fit into the American-led collective security framework. British naval exercises took place in the South China Sea on August 31, 2018. Gavin Williamson, who at the time was British Secretary of State for Defense, summarized the UK's South China Sea policies in a February 2019 speech: "We and our allies must deter and be ready to defend ourselves. Ready to show the high price of aggressive behavior. Ready to strengthen our resilience. And ready, where necessary, to use hard power to support our global interests. … We need to build on our established relationship with … Singapore and Malaysia in the Five

Powers Defense arrangement. With other ASEAN nations, with Japan, the Republic of Korea, and India."[174] While Britain wants to have a larger role in the South China Sea, they will have to focus more on their military to make the impact that they desire.

Out of all the European countries in recent history, France is the one most involved in Southeast Asia. David Scott notes that the United States has welcomed France's participation in naval exercises in the region, as Admiral Harry Harris said in 2018: "France has significant equities in the Indo-Pacific, and I welcome France's growing involvement in the region … France aims to become more involved across the Indo-Pacific." Scott elaborates that this cooperation could be seen earlier, when the French joined in exercises with America, and Japan in the western Pacific in May 2017.[175] France's move into the South China Sea can be fruitful for the QUAD, which will need all the help they could get in a war with China.

As France has territories in the Pacific, they have an interest in the Indo-Pacific. Morcos Pierre explained France's movement in the Indo-Pacific by writing that "Over the last few months, the French Navy's operational activity has been particularly intense in the Indo-Pacific."

---

174 Gordon, Houlden (Professor Gordon Houlden is Director Emeritus of the China Institute, Professor of Political Science and Adjunct Professor of the Alberta School of Business at the University of Alberta.), Romaniuk, Scott N. (Dr. Scott N. Romaniuk is a Newton International Fellow of the British Academy at the University of South Wales' Faculty of Life Sciences and Education, and a Visiting Fellow at the university's International Centre for Policing and Security), Hong, Nong (Executive Director of the Institute for China-America Studies and a Research Fellow at both the National Institute for South China Sea Studies, China, and the University of Alberta, Canada). *Security, Strategy, And Military Dynamics in the South China Sea: Cross-National Perspectives.* :Security, Strategy, and Military Dynamics in the South China Sea: Cross … - Google Books, Published by Bristol University Press. Bristol, Britain. 2021. 308-312, 316, 319.

175 Scott, David (A doctor presenter on Indo-Pacific geopolitics at the NATO Defense College in Rome and the Baltic Defence College in Tartu, Estonia). "US Indo-Pacific Strategy: The Indo-Pacific in US Strategy: Responding to Power Shifts - Rising Powers in Global Governance" (kureselcalismalar.com), *Rising Powers Quarterly* Volume 3, Issue 2, 2018, 19-43, Published by Marmara University. Istanbul, Turkey. August 2018. 34.

France describes itself as an "island state" in the region, with 1.6 million French civilians living in territories such as Mayotte, La Réunion, New Caledonia, and French Polynesia. These territories share borders with 17 nations in the Indo-Pacific. Pierre notes the territories' significance by writing: "More than one-third of French exports to non-EU countries go to the region. French direct investments amounted to 320 billion euros in 2018, a 75 percent increase since 2008. More than 7,000 French subsidiary companies are present in the region, with revenues increasing by 40 percent between 2010 and 2016." France has stationed 8,000 soldiers and dozens of ships in bases in the territories, accounting for 60 percent of France's military force overseas.

France has strengthened connections with India through arms deals. In a 2018 speech at the Garden Island military base in Sydney, President Emmanuel Macron called France an "Indo-Pacific power." France has provided military support for other nations in the Indo-Pacific, including India and Australia. Pierre writes that "Recent examples include India's decision to acquire 36 Rafale jet fighters or Australia's contract with French shipbuilder Naval Group to produce 12 attack submarines. French regular naval deployments in the region are also closely coordinated with its partners, as witnessed by aircraft carrier *Charles de Gaulle*'s mission in 2019 in which it conducted naval-air training operations with Japan, India, the United States, and Australia."

France is seen as a supporter of multilateralism and a pillar of stability. France is planning a long-term strategy for the Indo-Pacific, as "The (Biden) US administration mov(ed) full speed ahead toward the Indo-Pacific, there is now room for greater cooperation between Washington and Paris. Notwithstanding its rich defense cooperation with the United States, France has of late preferred to distance itself from US-China tensions, which it perceived as contributing to the polarization and instability of the region ... Given its long-standing presence in the Indo-Pacific, France could play an important role in US strategy by contributing to greater coherence of efforts from like-minded countries across the region. It has already started this with the France-India-Australia dialogue. France could also help drive EU policy toward the Indo-Pacific and ensure that Europeans play a valuable and coordinated role in the region."[176] France, being a major power in Europe, could encourage more continental powers to deter China from disrupting trade going through the Indo-Pacific.

While the EU has focused on internal matters since its creation, the increasing importance of the South China Sea and the growing tension between China and its Indo-Pacific neighbors are becoming too great to ignore. Loro Horta argues that with China's ambitions creating tension within the South China Sea, the EU is reconsidering its role in the region. In April of 2021, the 27 members of the European Union made a joint statement declaring that said tensions "endanger peace and stability in the region" and cautioning against "any unilateral actions that could undermine regional stability and the international rules-based order." The statement called for "all parties" to abide by the 2016 ruling by the Permanent Court of Arbitration that dismissed most of China's South China Sea claims. Beijing, notably, has not recognized the ruling. In that same month, the foreign ministers from the EU as the Council of the European Union, echoed the same sentiments as America, Japan, and Australia on the concerns on China. As the EU is the biggest foreign investor in ASEAN, "with stocks amounting in 2019 to more than €313.6 billion Euros (336.7 billion USD)" In turn, ASEAN is the EU's third largest trading partner, at €189 billion ($202.9 billion) in 2020.

Horta adds that "Britain, outside the EU, had been an early actor. Despite warnings from China, the British aircraft carrier HMS *Queen Elizabeth* and its escort vessels recently sailed the South China Sea. London also announced that it would permanently deploy two warships

---

176 Morcos, Pierre (a visiting fellow with the Europe, Russia, and Eurasia Program at the Center for Strategic and International Studies [CSIS] in Washington, D.C.). "France: A Bridge between Europe and the Indo-Pacific?" Center for Strategic and International Studies. April 1, 2021. https://www.csis.org/analysis/france-bridge-between-europe-and-indo-pacific.

in Asia." The French already have a presence in the South China Sea due to their Indo-Pacific territories. Their presence has increased, with a French nuclear submarine and support vessel passing through the Malacca Strait and visiting Vietnam in February.[177] Europe's shift to the Indo-Pacific shows a growing front that can protect free trade in the region and deter an aggressive nation from taking over trade in an ever-important region in the world.

Europe has already helped in terms of supporting the Philippines, which plays an important part in the South China Sea. Such help is elaborated by Jim Gomez, who writes that "'EU special envoy to the Indo-Pacific region Richard Tibbels said the 27-nation bloc is also ready to provide satellite surveillance to help countries like the Philippines respond to natural disasters and protect their interests, as tensions pitting China against its smaller neighbors escalate in the disputed waterway.'" Forty percent of foreign trade to the EU goes through the South China Sea. "We're talking to our like-minded partners. We want to be prepared. We want to work out what would need to be done in case of increasing tensions and that's quite a work going on behind the scenes," Tibbels said. "But I think you can obviously count on the reaction of like-minded partners being robust should untoward developments happen."[178] Europe's cooperation with the Philippines shows the gradual push to get involved in the South China Sea.

With the Russo-Ukrainian War looming large over the rest of Europe, the similarity of Russian aggression in Eastern Europe to Chinese aggression in the South China Sea is not lost on the EU.

---

177 Horta, Loro (deputy head of mission at the Timor Leste Embassy in Beijing. He is a graduate of the Chinese National Defense University senior officers course, the US Naval Post Graduate School, the American National Defense University and the S Rajaratnam School of International Studies.). "Europe and the South China Sea." *The Interpreter*. September 2, 2021. https://www.lowyinstitute.org/the-interpreter/europe-south-china-sea.

178 Gomez, Jim (Associated Press Chief Correspondent based in Manila, Philippines). "EU eyes more naval visits in disputed South China Sea." Associated Press. March 15, 2023. https://apnews.com/article/european-union-navy-south-china-sea-643ee2f29e52321fa9efaf481006f13c.

Basilio Sepe, JoJo Riñoza, and Noel Celis explore the actions of the EU in context of the Russo-Ukrainian War. The authors mention that with China spurring tension in the South China Sea, the EU has increased its contacts with the Philippines. On a visit to the Philippines, European Commission President, Ursula von der Leyen, agreed that the 2016 UNCLOS arbitration ruling was "legally binding." After meeting with the Filipino President Ferdinand Marcos Jr. in Manila, von der Leyen said Europe is ready to cooperate with the Philippines on maritime security. With the Russo-Ukrainian War going on, she said, "The illegal use of force cannot be tolerated. Not in Ukraine, not in the Indo-Pacific. Security in Europe and security in the Indo-Pacific (are) indivisible. Challenges to the rules-based order in our interconnected world affect all of us." Von der Leyen added that the Philippines is a great partner with the EU and will continue to be a key partner in the future. The EU has also commenced talks with Thailand, Singapore, Vietnam, Malaysia, and Indonesia, all nations that are a part of ASEAN.[179] Europe's investment in the Philippines could open the door for closer relations with ASEAN.

While Europe has relied on the United States and NATO for security in Europe, the EU is making its own strides to defend its interests in Southeast Asia. Emily Benson and Gloria Sicilia claim that the United States re-shifted focus on the Indo-Pacific in 2011, and the EU took similar action in 2016. Also in 2016, the EU announced its first official position on the Indo-Pacific with publication of European Union Global Strategy (EUGS). In it, the EU "recognizes the need to strengthen partnerships across the Indo-Pacific and East Asian regions to address common humanitarian, economic, and security challenges ... Since 2019, the European Union has negotiated several trade

179 Sepe, Basilio, Riñoza, Jojo, Celis, Noel (Contributors to BenarNews, an RFA-affiliated online news organization). "EU says it backs Manila in South China Sea dispute with Beijing." Radio Free Asia. July 31, 2023. https://www.rfa.org/english/news/southchinasea/philippines-china-eu-07312023141918.html.

agreements with the countries of Japan, New Zealand, Singapore, and South Korea."

With China supporting Russia in the Russo-Ukrainian War, the EU referenced China as a "systemic rival." The EUGS strategy paper also contained a thinly veiled warning that China should not give Russia any stronger support: "The European Union should … leverage its economic and diplomatic influence to ease regional tensions and strengthen ties with Indo-Pacific nations sharing economic and security interests amid growing Sino-American rivalry."[180] As Europe strengthens its ties with Asia, its rivalry with China becomes more clear.

As China is helping Russia in the war against Ukraine, the EU has been exploring ways to distance themselves from China's economic and political power. This is explained by Suzanne Lynch, who mentions that before Biden's 2023 meeting with von der Leyen, American officials tried to warn the EU about China and its promises to resolve trade disputes. Recently, America has given intelligence to the EU that China is considering arming Russia in their war with Ukraine. German Chancellor Olaf Scholz warned China against doing so, but von der Leyen has taken a "more careful" approach. Heather Conley, a former State Department official overseeing European and Eurasian Affairs in the George W. Bush administration said, "The Europeans have already experienced deep economic trauma because of cutting off Russia. They cannot imagine cutting off China."

In her 2023 comments, von der Leyen said both Germany and the Netherlands were re-evaluating some trade with Chinese high-tech firms in high-tech items. She noted that "China produces 98 percent

180 Benson, Emily (the director of the Project on Trade and Technology and senior fellow with the Scholl Chair in International Business at the CSIS in Washington, D.C.), Sicilia, Gloria (an intern with the Project on Trade and Technology at CSIS). "Navigating Tides: The European Union's Expanding Role in the Indo-Pacific." CSIS. March 5, 2024. https://www.csis.org/analysis/navigating-tides-european-unions-expanding-role-indo-pacific.

of Europe's supplies of rare earths (and) … Europe needs to de-risk this dependency." She added that "the EU wants to 'de-risk' but not 'decouple' from China."[181]

Though Europe wants to distance themselves from China, it would be too costly for them to completely cut off their major trading partner.

While Europe is far from Southeast Asia, their investment in the region makes it difficult for them to ignore the growing tension in the region. Britain, France, and Germany especially have an interest in securing a lasting peace in the region. Though Europe can play a supportive role in the South China Sea War, it would be preferable for the nations on the continent to mediate a stronger peace in the South China Sea and get America and China to agree to fewer military exercises in the region. If Europe wants to form an everlasting link with Southeast Asia, it must work with the other nations to form a more perfect tranquility in the region.

---

181 Lynch, Suzanne (POLITICO's Global Playbook Author. She joined POLITICO Europe after more than four years as the Washington Correspondent for the Irish Times), Toosi, Nahal (POLITICO's senior foreign affairs correspondent), Moens, Barbara (Chief EU Correspondent at POLITICO, based in Brussels. Prior to joining POLITICO, Barbara covered Belgian politics for the Belgian national daily *De Tijd*), Banco, Erin (a national security reporter for POLITICO focusing on the intelligence community). "The U.S. wants Europe to stand up to China. Europe says: Not so fast." Politico. March 8, 2023. https://www.politico.com/news/2023/03/08/us-europe-china-00086204.

Chapter 15: Two Scenarios if There Is a War

Ever since the end of the Vietnam War, the countries of Southeast Asia have been preparing for the next major war in the region. The Chinese government's saber-rattling has caused their neighbors to arm themselves to defend their territory from this aggression. Taiwan remains the main target for Chinese war plans, which has made the island nation a rallying point for other major nations in the South China Sea to counter China's ambitions to take over the South China Sea. This has also resulted with China using its economic leverage to get support from other countries. It will be important to explore how both sides of the conflict will fight on the battlefield to prepare for the worst-case scenario.

While China has plans to take Taiwan, it is important to figure out what their strategy will be on the Taiwanese front of a South China Sea War. Evaluations have been made to what China could do to Taiwan in a war. Robert D. Blackwill and Philip Zelikow predict that China could show force by invading one of Taiwan's offshore islands, like the Taiping Island, the Pratas Islands, or the Penghu Islands. They write that "Of these four kinds of choices, Pratas would seem to be the most tempting for Beijing." An attack on the offshore islands can be seen as a low-risk strategy in comparison to invading Taiwan proper. However, this would spur the United States and Japan into action and could land two of the strongest nations in Southeast Asia into a fight against China.

An alternative would be for China to blockade Taiwan by taking control of their air and sea borders with more of a military presence in said borders. China might defend this action by claiming it was simply establishing Chinese sovereignty over Taiwan. A blockade could also stop American arms exports going into Taiwan. This could bring nations like America and Japan to the negotiation table. This could also cause America and Japan to do a counter blockade in China.

Alternatively, China could do a traditional amphibious invasion in Taiwan, or an airborne assault with help by special operatives. "Such assaults are risky and difficult operations," Blackwill and Zelikow state. "The Taiwanese would have some significant inherent advantages. The attacker could only get a small number of forces ashore in an initial assault. The defender should be able ... to amass much larger forces in the vicinity of the assault unless those forces can somehow be deflected and paralyzed by the attacker ... the invader would need to sustain the forces that have made it ashore with the thousands of tons of supplies needed to maintain modern military operations." The weather in the Taiwan Strait would make an amphibious invasion difficult in most of the year. The terrain of Taiwan also presents difficulties for such an invasion plan. With Taiwan investing in missiles and sensors in the past 10 years, they can do more damage to China with cheaper weapons, while doing damage to Chinese ships and aircraft.[182] These options are what China could use and what Taiwan, America, and their partners need to look out for.

While China has planned to take back Taiwan since 1949, the geography of the island has been the main problem for the invasion plan. David Sacks claims that with Taiwan creating its own identity apart from China and its reluctance to peacefully unite with the mainland, China could be driven to go to war. China would have to deal with the 90-mile-wide Taiwan Strait, where two monsoon seasons and extreme weather appear to make a naval invasion viable for only

---

182 Blackwill, Robert D. (the Henry A. Kissinger senior fellow for U.S. foreign policy at the Council on Foreign Relations), Zelikow, Philip (the White Burkett Miller professor of history and Wilson Newman professor of governance at the Miller Center, both at the University of Virginia). "The United States, China, and Taiwan: A Strategy to Prevent War." Council Special Report No. 90 February 2021. Published by the Council of Foreign Relations in New York, New York. 32-38.

a few months out of the year. Transporting the hundreds of thousands of troops needed for the invasion would take weeks, and thousands of ships would be required. With each crossing taking hours, Taiwan would be able to target troops and prepare defenses.

The number of Chinese amphibious vehicles is relatively low and though China could use civilian vessels, it would take longer for the troops to unload and would be vulnerable to Taiwanese missiles. There are few deep-water ports and beaches for China to land a large landing force. Taiwan could destroy these ports to prevent China from using them. The west coast of Taiwan has shallow waters, which makes it hard for China to invade with a large landing force. Taiwan's east coast has cliffs that are too steep for an invasion force. Sacks emphasized the difficulty of a Chinese invasion of Taiwan by comparing it to the D-Day operation, which is the biggest naval invasion in history, with 850,000 troops landing in Normandy over the course of three weeks. He speculates that China's invasion of Taiwan would require bigger numbers than that. It does not help that Taiwan has invested in anti-landing spikes, mines and mobile missile launchers. Taiwan has taken part in anti-landing exercises and tested out naval mines during military drills. Despite China's modernization of its military, Taiwan's inherent advantages would be difficult for China to overcome. Even trying to take Taipei, Taiwan's capital, would prove difficult as the city sits in a bowl of mountains that defenders can use to counter China. Taiwan can also destroy Taipei's major port, tunnels and highways leading to the city to prevent China's seizure of the city. Taiwan's terrain makes it so that the populace lives in a handful of cities, with 7 million in Taipei. This means that China would have to fight in urban combat against the Taiwanese people as well.[183] An invasion of Taiwan is a plan that would either be a stroke of brilliance or a monumental disaster, which explains China's caution on the issue.

---

183 Sacks, David. "Why China Would Struggle to Invade Taiwan." Council on Foreign Relations. January 10, 2024. New York, New York. https://www.cfr.org/article/why-china-would-struggle-invade-taiwan.

Given their proximity to each other, China and Japan will be on the frontlines in the event of a war. The scenario where China and Japan go to war with each other is explored by Dr. Gary Bowman, Dr. Fabio Caccioli, Dr. Scott Kelly, Dr. Roxane Foulser-Piggott, Daniel Ralph, and Simon Ruffle. They estimate that a war between Japan and China could lead to the death of up to 500,000 people, with millions more injured due to strategic bombings. They also predict infrastructure damage totaling up to $500 billion. This would lead to a loss of GDP output for five years, resulting in economic losses of $17-$34 trillion. Not only would oil and energy prices rise during the war, Chinese and Japanese exports would drop by $1.5 trillion and $726 billion, respectively.[184] A war between China and Japan alone would be a disaster for the South China Sea, as the crossfire would stop trade and would create an international crisis.

One can also look at history to estimate how many people would perish in a South China Sea War. Nikki Sway estimates that over the course of several decades, 47,000 people have lost their lives because of the conflicts involving the Kashmir region. Despite several ceasefires throughout the years, tensions between India and Pakistan over Kashmir still rise to the point of small skirmishes.[185] A war that brings in India and Pakistan would escalate what is already occurring in Kashmir and potentially form a new theater of war between India, China, and Pakistan.

---

184 Bowman, Dr. Gary (Geopolitical Conflict Project Lead), Caccioli, Dr. Fabio (Research Associate), Kelly, Dr. Scott (Research Associate), Foulser-Piggott, Dr. Roxane (Research Associate), Coburn, Dr. Andrew (Director of External Advisory Board, Professor), Ralph, Daniel (Academic Director), Ruffle, Simon (Director of Technology Research). "China-Japan Geopolitical Conflict Scenario" :China-Japan Geopolitical Conflict Scenario (drscottkelly.com). Cambridge Centre for Risk Studies: Cambridge Risk Framework: Geopolitical Conflict: Stress Test Scenario. Published by the Cambridge Centre for Risk Studies, University of Cambridge Judge Business School. Cambridge, Britain. June 2014. 3, 38.

185 Sway, Nikki (Senior at Bard College). "India's Military and its Impact on the Formation of a Nation." :India's Military and its Impact on the Formation of a Nation (bard.edu). Bard Masters of Art in Teaching Program Academic Research Project. Published by Bard College. Annandale-On-Hudson, NY. January 28, 2020. 8.

What is particularly worrisome about a war between India and Pakistan is that both nations have nuclear weapons. Christopher Clary and Vipin Narang cite the book *Choices* by Shivshankar Menon, an Indian diplomat and former national security adviser. In the book, Menon speculates about the possible use of nuclear weapons: "There is a potential gray area as to when India would use nuclear weapons first against another NWS (nuclear weapons state). Circumstances are conceivable in which India might find it useful to strike first, for instance, against an NWS that had declared it would certainly use its weapons, and if India were certain that adversary's launch was imminent."

India's strike-first policy can be seen as far back as 2000, when then Indian Prime Minister Atal Vajpayee said, "We are being threatened (by Pakistan) with a nuclear attack. Do they understand what it means? If they think we would wait for them to drop a bomb and face destruction, they are mistaken." In 2015, then Indian Lieutenant General Nagal said, "Once the adversary has decided to initiate nuclear attacks on India or Indian forces anywhere, the aim (of India) will be to terminate the war at the earliest." India has also developed cruise missiles such as the BrahMos and Nirbhay, the former being a joint measure with Russia, as a "nuclear-based" system. "Fired from land, air and sea platforms, [the BrahMos] can hit targets at ranges of 300 to 500 kilometers and is also capable of carrying a nuclear warhead." The Nirbhay can go 1,000 km and can carry a 300 kg warhead. In response to India's expansion of nuclear weaponry, Pakistan has built up infrastructure to engage in an arms race with India. It is estimated that in 2025, Pakistan could have as many as 250 nuclear warheads.[186] Whatever India's strategy is when

186 Clary, Christopher (Christopher Clary is an Associate Professor of Political Science at the University of Albany.), Narang, Vipin (Vipin Narang is the Principal Deputy Assistant Secretary of Defense for Space Policy. Narang has been an American political scientist. He is Professor of Political Science at MIT.). "India's Counterforce Temptations: Strategic Dilemmas, Doctrine, and Capabilities" :India's Counterforce Temptations: Strategic Dilemmas, Doctrine, and Capabilities | International Security | MIT Press. International Security (2019) Volume 43 Issue 3: 7–52. Published by

countering Pakistan, is they are willing to use nuclear weapons when going to war with Pakistan.

A war fought for Kashmir between India and Pakistan would exacerbate the atrocities that are already happening in the region. Said atrocities are covered by Saghir Iqbal, who alleges that there have been cases of genocide and mass graves within Kashmir. He cites analyst Riaz Haq, who said: "The essence of Kashmir issue today is not Uri or Pathankot or similar other alleged 'militant attacks;' it is India's brutal military occupation of 700,000 heavily armed Indian soldiers being resisted by 10 million Kashmiris. Anyone who tells you otherwise is a liar." Dr. Angana Chatterji, a professor of cultural and social anthropology at the California Centre for integral studies, said that "violence and militarization in Kashmir, between 1989-2009, have resulted in over 70,000 deaths, including through extrajudicial or fake encounter executions, custodial brutality, and other means." Since January 1989, over 10,000 Kashmiri women have reportedly been raped by the Indian army. Iqbal estimates that from January 1989 to March 2016, over 94,000 Kashmiris were killed."[187] Though the estimates of past death and destruction vary widely, it is clear that the blood already spilled in Kashmir offers a dark prediction of what could happen if war is to be waged there.

With India and Pakistan investing in their influence in Kashmir, the on-going proxy war in the region could foreshadow what an all-out war would look like, not only in Kashmir, but any land theatre in the South China Sea War. Said proxy war is covered by Werner Menski and Muneeb Yousuf. The authors start by mentioning that

---

MIT Press. Cambridge, Massachusetts, U.S.A. February 1, 2019. 17-18, 21-22, 29-31, 39-40, 48.

187 Iqbal, Saghir (a researcher in International Relations and Security Studies in the University of Lynchburg in Lynchburg, Virginia). *Pakistan's War Machine: An Encyclopedia of its Weapons, Strategy and Military Security* 2018 1st Edition. :Pakistan's War Machine: An Encyclopedia of its Weapons, Strategy and … - Saghir Iqbal - Google Books. Published by CreateSpace Independent Publishing Platform. Scotts Valley, California, U.S.A. April 13, 2018. 21-23, 25, 27-29.

in 2019, the Indian government abolished the special constitutional status of Jammu and Kashmir to incorporate the region as a union territory in India. This move led to protest and violence in the region. "One major characteristic of this tragedy remains that J&K's (Jammu and Kashmir) people themselves are not the main actors but certainly the major victims," Menski and Yousuf write. Pakistan's support of Jihadists in the Kashmir region has backfired and proves to be a thorn in Pakistan's side. India tried to stop the jihadist violence but ended up being blamed for "excessively aggressive responses." It does not help that a 2019 report from the UN showed that India and Pakistan had committed human rights violations on the Kashmiris. On the Pakistan side of Kashmir, individuals, families and communities became victims of state-backed terrorism, and religious terrorists who claim the moral high ground as "defenders of Islam." Menski and Yousuf conclude that "Kashmiris appear to be in the same leaky boat as Kurds, Palestinians, and other communities in the world, trying to survive decently at the edge of large nation states, while asking for a fair deal."[188] This leaves Kashmir in an unwinnable position if war were to erupt between India and Pakistan.

All this is made worse, because a war between India and Pakistan can become nuclear. Iqbal comes back to cite former soldier and UN Peacekeeper Raghu Raman, who said, "For all the chauvinistic war mongering touted in every medium, India cannot 'win' a war against Pakistan and the sooner we appreciate this politico-military reality, the more coherent and serious we will sound to our adversaries and the world community ... Matter of fact, short of total genocide, no country regardless of its war-withal can hope to achieve a decisive victory with a 'short war' in today's world. As the US is discovering eight years,

---

188 Menski, Werner (Member of the SOAS South Asia Institute in London, Britain), Yousuf, Muneeb (a Kashmiri scholar who completed his master's in conflict analysis and peace-building from the Nelson Mandela Centre for Peace and Conflict Resolution, Jamia Millia Islamia University, New Delhi). *Kashmir after 2019: Completing the Partition.* :Kashmir after 2019: Completing the Partition - Google Books. Published by SAGE Publications India. New Delhi, India. 2022.

trillion dollars and over 25,000 casualties later—in Afghanistan. That era of 'decisive' short wars—Zix especially in context of an Indo-Pak war is largely over because of several reasons." If nuclear weapons were to hit the five biggest cities in India and Pakistan, the total of deaths in India would be 1,690,702 people, and for Pakistan it would be 1,171,879 people.[189] The Indo-Pakistani war would be one of the bloodiest theaters of war in a South China Sea War, second only to the Taiwanese theater.

While there are risks to the invasion of Taiwan, that risk is worth taking. China, Pakistan, North Korea, and to a lesser extent Russia, could form a military alliance to fight against America and its allies in the South China Sea. If successful, China and its partners would shift the power balance in the region and take one more step for China to become the world's superpower. By winning the war, China would occupy Taiwan and their claims in India, and have their claims over the entirety of the South China Sea recognized by the world. Pakistan could own the Kashmir region and hinder India's attempt to become a great power. North Korea could take over South Korea and reunify the Korean Peninsula under the Kim Dynasty. Russia could get Japan to recognize its ownership of the Kuril Island and rekindle far east Asian ambitions. China would be at the forefront of a new world order and would leave America and its allies in uncertainty and disillusionment.

Although China is adamant in its attempt to take over the South China Sea, America is equally determined to fight against the new rising world power. The QUAD organization, South Korea, Europe and ASEAN have something to gain from an American victory and the reinforcement of the international order. An American victory over China would be the first over a major world power since World War II. It would give the United States breathing room on the world stage, as no other major power could threaten them. India would get full control over the Kashmir and Ladakh regions, and could help create a buffer state in Tibet to prevent another direct war with China. Japan could

189 Menski and Yousuf. 30, 42.

settle the dispute with Russia over the Kuril Islands. Australia could break free from the influence that China has put onto the country. South Korea could defeat the North, and unite the Korean Peninsula under a democratic government. Europe would be able to assure the safety of its Indo-Pacific civilians and create a better relationship with the nations of the South China Sea. ASEAN could bring China to the talking table and settle the territorial disputes over the South China Sea. China and its allies would be in devastating waters as they struggled to salvage what was left of the nations and the resolve of their people.

While war will always have the victors and the defeated, it is important to remember that a war on such scale would mean a catastrophic end to the peace and prosperity in the South China Sea. Millions of human beings would die and even more would suffer from injury. Billions of dollars' worth of damage would be caused by the conflict, destroying decades of infrastructure. Given that trade acts as the arteries of the nations of the South China Sea, a major war in the region would be like a bowling ball falling through a spider's web. While America and China are preparing for war, it is important that war remains the last option for all parties involved.

Chapter 16:
How to Stop Escalation

While the possibility of military conflict in the South China Sea remains a significant concern, it is important to remember why a major war has not happened in the region since the Vietnam War (1955-1975). The economic breakdown that would come from the war would not just affect the South China Sea nations, which rely on the sea for energy and food imports, but would generate ripples throughout the world as most electronics and consumer goods come from the South China Sea region. Another thing to note is that the buildup of military might in the region raises the human cost of sparking a major conflict, as military weaponry becomes more deadly by the decade. With most powers in the South China Sea relying on China for their economy and America for their military, the political backlash from a war would force the nations in the region to choose between two of the biggest powers in the world. It is pertinent to remind the nations of the South China Sea the dangers of war before national leaders threaten large-scale violence on their neighbors.

Given their rocky relationship, the United States and China have a complex history. Said history is explored by Thomas J. Christensen, who starts by showing how America's open markets helped China economically in the late 1970s. Chinese exports were able to access the United States market, and hundreds of thousands of Chinese students were able to be taught in American schools. This also allowed the United States to do large scale investments into Chinese industry. All of this helped China to modernize and to grow at a fast rate technologically. Christensen mentions, however, that the United States has tried to stop the growth of China's influence on the international stage. The Tiananmen Square Massacre, on June 4, 1989, led to the United States and its European allies putting an arms embargo on China, one that stands to this day. Christensen writes that "The current US-China strategic competition, which is real and carries dangers of its own, lacks three essential and interrelated elements of the US-Soviet Cold War: the United States and China are not involved in a global ideological struggle for the hearts and minds of third countries; our highly globalized world is not and cannot be divided into starkly separated economic blocs; and the United States and China are not leading the kind of opposing alliance systems that were the precursors of the bloody proxy wars of the 20th century."

He writes that, with the exception of Taiwan, the islands and reefs that are in China's scope are not tempting targets of invasion as they would be exposed to enemy fire. Even the artificial islands that China has been making since 2014 would be vulnerable to the same problems if China were in a war. Christensen cites former Chinese diplomat to Russia Shi Ze who argues that, unlike Russia, China wants to work within the world order: "China and Russia have different attitudes. Russia wants to break the current international order … Russia thinks it is the victim of the current international system, in which its economy and its society do not develop. But China benefits from the current international system. We want to improve and modify it, not to break it."

Christensen adds that "The United States enjoys massive superiority to China in one key realm of national security: allies and partnerships. China's small number of active partners constitute a rogues' gallery of relatively weak and sometimes volatile actors with limited ability to enhance China's power around the world. The United States network of over 60 global allies and security partners include many of the most advanced, high-tech economies in the world, including Japan, South Korea, Australia, Germany, the United Kingdom, France, Israel, Singapore, etc. This US-led security network is what allows the United States the power projection necessary to be a truly global superpower. Many US partners would likely side with the United States when a rising China is to become very aggressive and expansionist. Chinese elites almost certainly know this; and that is one of the many reasons that a rising PRC, despite multiple sovereignty disputes with its neighbors, has remained relatively restrained (before the recent Sino-Indian border skirmishes China had not been in a military conflict since 1988 and China has not been in a true war since the Sino-Vietnamese War of 1979)."[190] Though America and China do not want an open conflict between each other, the tension between the two could create an opening for the United States to gain help from its alliances and exploit China's military inexperience.

The internal problems of China could be a deciding factor for whether China goes to war with Taiwan. Alicia Garcia-Herrero and Robin Schindowski begin by mentioning that 20 percent of Chinese youth are unemployed after the reopening of the economy after COVID-19. Despite decades of growth, China is still dependent on foreign technology for sectors such as semiconductors. They explain this by writing that "Draconian competition and a lack of social mobility might become a problem for innovation. One of the major issues at hand is that China is churning out university graduates at a faster pace than the economy can absorb. This has led to ballooning youth unemployment in the cities, reaching 20 percent in 2022 and still not dropping despite the full reopening of the economy ... Another sign of this youth fatigue has been the 'flat' movement of 2021 in which Chinese netizens celebrated sloth and disinclination to exertion." They emphasize that China is also dealing with a rapidly aging population,

---

190 Kaplan, Robert D. (bestselling author on foreign affairs and holds the Robert Strausz-Hupé Chair in Geopolitics at the Foreign Policy Research Institute. He was a member of the Pentagon's Defense Policy Board and the U. S. Navy's Executive Panel.). *Asia's Cauldron: The South China Sea and the End of a Stable Pacific* Asia's Cauldron: The South China Sea and the End of a Stable Pacific - Robert D. Kaplan - Google Books. Published by Random House LLC., Manhattan, New York. U.S.A. 2014. 7, 9-10.

which reduces the labor force.[191] China's problem with dissatisfied youth could prevent China from fighting at their full capacity, as more young people grow disillusioned with the Chinese government.

One way that the war can be prevented is to make the countries of the South China Sea understand how devastating a war would be on their economies. Economic interdependence in the region is covered by Rebecca Strating. Strating cites American National Security Advisor Robert O'Brien who offers one example of war's impact by saying that China could "try and stop ASEAN [Association of Southeast Asian Nations] nations from exploiting the offshore resources, blocking access to $2.5 trillion of oil and gas reserves alone." The Center for Strategic and International Studies (CSIS) estimated that, in 2016, $3.37 trillion worth of trade went through the South China Sea. With that in mind, a blockade from China would result in an energy security and economic crisis. Japan is a prime example as former Prime Minister Abe said: "China is an indispensable country for the Japanese economy to keep growing. We need to use some wisdom so that political problems will not develop and affect economic issues."[192] To prevent this war, the countries of the South China Sea need to realize how prosperous they have become because of the relative peace that they have attained since the end of the Vietnam War.

It is also important for nations in the South China Sea to weigh the benefits of having ties to both America and China before a war makes them side with one of the two major world powers. John G. Ikenberry discusses the benefits China and America have as they partner with the

---

191 Garcia-Herrero, Alicia (Senior Fellow at Bruegel), Schindowski, Robin (A Research Assistant at Bruegel). "China's Quest For Innovation: Progress And Bottlenecks." Working Paper Issue 08/2023. Published by Bruegel in Brussels, Belgium. June 21, 2023. 1-2, 19-20, 21.

192 Strating, Rebecca (Senior Lecturer in Politics and International Relations in the Department of Politics, Media and Philosophy at La Trobe University, Melbourne, Australia). "Defending the Maritime Rules-Based Order: Regional Responses to the South China Sea Disputes.": Defending the maritime rules-based order: regional responses to the South China Sea disputes (hawaii.edu). Policy Studies|80. Published by the East-West Center. Honolulu, Hawaii. 2020. 8, 13, 27-28.

other nations in the South China Sea: "In this hegemonic competition, the United States and China can both offer the benefits of their leadership. They will each have carrots and sticks to push and pull the region in their direction. The United States has its extended security system as an asset, and China has its trade and investment assets. Each state will seek to offer the region its hegemonic services. Each state also has tools with which to apply political pressure. But there are costs and unwelcome consequences if the United States and China attempt to use, respectively, their security and economic leverage. In the end, it will be the middle states that have the ability to shift the regional order in one direction or the other—or to make choices to preserve the double hierarchy."[193] There are economic and security benefits that China and America respectively give to the nations they partner with in the South China Sea. This creates incentives for the other nations to avoid putting the region in the position of a war between China and the United States, as such a conflict would undermine the economy and security of the region.

The hope for long-term peace for the South China Sea can only come from America and China reaching a settlement about their grievances in the region. Jiacheng Li suggests that China, America and other powers should come to talks about each nation's rights and privileges in the Indian Ocean. That could be much like the Montreux Convention Regarding the Regime of the Straits in 1936, in which Turkey, Greece, the Soviet Union, France, and the United Kingdom reached agreements about how these nations and their ships should behave in the Bosporus Strait and the Black Sea.[194] Such a settlement

---

193 Ikenberry, John G. (a theorist of international relations and United States foreign policy, and the Albert G. Milbank Professor of Politics and International Affairs at Princeton University). "Between the Eagle and the Dragon: America, China, and Middle State Strategies in East Asia." : Between the Eagle and the Dragon: America, China, and Middle State Strategies in East Asia on JSTOR. *Political Science Quarterly* Vol. 131, No. 1 (Spring 2016), pp. 9-43. Published by Oxford University Press. Oxford, Britain. 2016. 11.

194 Li, Jiacheng (an associate professor at Liaoning University's School of International Studies in Liaoning, China). "China's Indian Ocean Strategy:

can keep the South China Sea at peace, but it would require both America and China to be open to compromise and maintaining an agreed-upon status quo in the region.

A South China Sea War can be avoided by reminding the nations of the South China Sea of the consequences that a major war would have in the region. Robert D. Kaplan explains that a major war in the South China Sea can be avoided, thanks to what is known as the "stopping power of water" theorized by University of Chicago political scientist John J. Mearsheimer. Mearsheimer explains that it will be harder for an army to commit a naval invasion than a continental invasion. The Taiwan Strait is a hundred miles wide—four times wider than the English Channel the Allies crossed to storm Normandy in WWII. China could take Taiwan, yet the geography of the nation would make it hard to occupy. A third of all maritime trade goes through the Malacca, Sunda, Lombok, and Makassar Straits. Two-thirds of South Korea's energy supplies, 60 percent of Japan's and Taiwan's energy supplies, and 80 percent of China's crude-oil imports come through the South China Sea. South China Sea contains more oil than most of the world except for Saudi Arabia.[195] The trade disruption caused by a war in the South China Sea would affect all nations in the region. Trade deficits could be especially problematic if China attacked Taiwan but was unable to achieve a decisive victory. That could make China reconsider going into a war that they cannot pull out of.

While Europe would have a limited military role in a South China Sea War, their diplomatic role could be a deciding factor in the war.

Developing China's Indian Ocean Strategy: Rationale and Prospects." *China Quarterly of International Strategic Studies* Vol. 03, No. 04, pp. 481-497 (2017). Published by World Scientific Publishing Co Pte Ltd. Singapore. 497.

195 Kaplan, Robert D. (bestselling author on foreign affairs and holds the Robert Strausz-Hupé Chair in Geopolitics at the Foreign Policy Research Institute. He was a member of the Pentagon's Defense Policy Board and the U. S. Navy's Executive Panel.). *Asia's Cauldron: The South China Sea and the End of a Stable Pacific* Asia's Cauldron: The South China Sea and the End of a Stable Pacific - Robert D. Kaplan - Google Books. Published by Random House LLC., Manhattan, New York. U.S.A. 2014. 7, 9-10.

Sarah Kirchberger, Svenja Sinjen, and Nils Wörmer explain that both France and Germany have tried to convince Russia to refrain from being dependent of China, by joining forces with the rest of Europe. However, China's support of the Putin regime has led Russia to a closer partnership with the Chinese. Europe's attempt to divide Russia and China is not only a matter of politics, but also of trade. The authors explain the importance of Europe-China trade by writing: "The volume of trade between China and Germany today almost equals the volume of Germany's trade with the US and France."[196] For now, Russia's war with Ukraine causes the Kremlin to focus more on western Russia than the Far East. In the longer term, Europe and NATO at large need to drive a wedge through Russo-Chinese relations.

While America and China are rivals on the South China Sea and the world stage at large, this does not mean that it must be a bloody rivalry. Scott Kennedy, a senior adviser and trustee chair in Chinese business and economics at the CSIS, elaborates on the idea in the article "US-China Relations in 2024: Managing Competition without Conflict." Kennedy starts by setting the scene, "… from 2018 to 2023, US-China relations were in a linear downward spiral. The trade war, the pandemic, growing technology competition, rising tensions in the South China Sea and the Taiwan Strait, and contrasting approaches to the Russia-Ukraine conflict have collectively fed a sense of fatalism that the countries were heading toward the abyss of outright economic decoupling and a disastrous military conflict."

He explains that the summit between Xi and former US President Biden in November of 2022 was helped by a year-long calming of

---

196 Kirchberger, Sarah (the Head of Asia-Pacific Strategy & Security at ISPK. She also serves as Vice President of the German Maritime Institute [DMI] and is a Nonresident Senior Fellow at the Atlantic Council), Sinjen, Svenja (is head of science communication at the Foundation for Science and Democracy [SWuD] and an editor at the journal *SIRIUS*), Wörmer, Nils (is head of the Konrad-Adenauer-Stiftung's International and Security Affairs Department since September 2018). *Russia-China Relations Emerging Alliance or Eternal Rivals?* Published by Springer. New York City, New York. June 16, 2022. 279, 289.

the waters. Kennedy argues that while America and China are in competition, the likelihood that they go to war with each other has gone down. Although China and America share different positions on Taiwan, Ukraine, human rights, technology, and economic security, Kennedy writes that the G-20 summit led to resumption of military-to-military talks, an agreement to expand renewable energy, and lower carbon emissions. Both the American and Chinese economies are embedded with each other with interdependence. "The new normal of what could be called 'competition without conflict' does not presage a return to the era of engagement," Kennedy wrote, "but the United States and China deserve credit for reducing the chances of the worst negative tail risks, including outright decoupling and military conflict. That said, the short-term incentives and long-term structural features of relations guarantee absolutely nothing. Maintaining momentum in 2024 will require a series of steps on the domestic, bilateral, and multilateral fronts."[197] America and China could compete in a peaceful manner that would avoid conflict.

"The new normal" however, has been halted as America and China are still at odds with each other over the South China Sea. David C. Gompert, Hans Binnendijk, and Bonny Lin claim that while China and the United States have cooperated in the fields of finance, development, nuclear proliferation, and counterterrorism, the tensions in East Asia have intensified their rivalry. As the authors state, "Because war could be catastrophic, both powers are strongly inhibited from starting one. Objectively, neither side stands to gain as much as it could lose from war, even if it does not lose militarily. Thus, war between China and the United States would most likely begin by a misjudgment or

197 Kennedy, Scott (senior adviser and Trustee Chair in Chinese Business and Economics at the Center for Strategic and International Studies [CSIS]. A leading authority on Chinese economic policy and U.S.-China commercial relations, Kennedy has traveled to China for 36 years.). "U.S.-China Relations in 2024: Managing Competition without Conflict." Center for Strategic and International Studies. January 3, 2024. https://www.csis.org/analysis/us-china-relations-2024-managing-competition-without-conflict.

miscalculation by one or both—by a misjudgment or miscalculation so consequential that it would qualify as a blunder (as we have defined it)."

The United States's relationship with Japan, Vietnam and the Philippines has given China the sense of paranoia about being encircled. The authors add that to the United States, the Chinese strategy appears to be an effort to diminish American presence in the Western Pacific. The authors explain that, since both America and China have interests in the Korean Peninsula, they would want to avoid going to war to maintain said interests. Strategic distrust between America and China could lead them to go to war with each other. However, neither Chinese or American decision makers have the tendency to make a blunder into war. The authors argue that, due to the devastation inherent in a war between the United States and China, a major conflict between the two powers remains unlikely. However, the chance of war can be further diminished if the United States and China foster strategic trust with one another by engaging in further contacts with each other's militaries.[198] Although America and China do not want to declare war on each other, they are far from strengthening ties with each other.

Given America's wealth and military capabilities, it has several potential ways to deter China from going to war against Taiwan. Such methods are explored by Oriana Mastro, who starts by arguing that the United States can use its military presence in the South China Sea as a deterrence to China in the form of a threat of punishment. The US can make the case that China would not succeed in a war in the South China Sea. He adds that America could bring China and the other claimant nations to the negotiating table and persuade China that they would not need to use military action to achieve their goals. America will have to be open with its grievances about China, which

---

198 Gompert, David C., Binnendijk, Hans, Lin, Bonny (Members of the RAND Corporation). "Chapter Seventeen the Sino-U.S. Case." *Blinders, Blunders, and Wars: What America and China Can Learn.* Published by RAND Corporation in Santa Monica, California. 2014, 217-218, 223, 225, 228-229, 239, 244-245.

can open up other nations to criticize China. Mastro writes that "to deter China, the US Congress could require the secretary of defense to report on any Chinese activities that infringe on the rights of other countries, such as swarming Filipino fishing vessels, to show China that aggression would incur an international reputational cost … The United States could also sanction Chinese companies and individuals involved in illegal activities, harassment, and coercion in the South China Sea."

Mastro suggests that America should encourage nations in ASEAN to conduct joint military exercises. With the help of other nations, America can enforce a code of conduct in the South China Sea, he contends. America could also negotiate with China to reduce military forces. This would be in exchange for America operating farther from China's island outposts and for China to demilitarize those islands.[199] The United States could use its current assets to get China to negotiate its claims in the South China Sea and the future of the region.

America will need to look back into recent history to find a solution to its rivalry with China. Andrew Faulhaber discusses possible ways to avoid a war in the South China Sea in a 2023 article. He states that "… the US military needs a strategy that not only promotes vital US national interests but also relays the credible use of force to deter Chinese aggression and is sustainable over the long-haul of strategic competition with China." He adds that the United States and China need more cooperation in intelligence and cybersecurity, as China looks at using both subjects as tools in a future war. Faulhaber goes further to argue that the conflicts in Vietnam, Afghanistan, and Ukraine should have taught the US that conventional military tactics are not enough to win a modern war.

199 Mastro, Oriana (Assistant Professor of Security Studies, Georgetown University's Edmund A. Walsh School of Foreign Service). "Military Confrontation in the South China Sea." Contingency Planning Memorandum No. 36. Published by the Council on Foreign Relations. May 21, 2020.

These conflicts show that a smaller military force can survive against a larger one with asymmetric military tactics, weapons procurement, and better strategy. American support of the Mujahideen in the Soviet Afghan War (1979-1989) and of Ukraine during the ongoing Russo-Ukrainian War show that by providing advanced weapons to its allies, the United States can achieve geopolitical strategy with little cost to the economy and people of America, in comparison to a traditional military conflict. Applying the same strategy to Taiwan could help deter China from an invasion. Faulhaber argues that the United States needs to create an Indo-Pacific NATO equivalent to fight Chinese expansionism. There are templates for such an organization with the Quad and AUKUS, which were formed due to deteriorating relations between China and the nations of the Indo-Pacific. With proper support from the United States, a NATO-like organization in the Indo-Pacific could deter China from expanding outside its borders, according to Faulhaber: "Just as China recently dictated to South Korea their 'three Nos' for South Korea's defense policies and weapons procurement if they do not want to face consequences, so too ... the United States (should) dictate their 'Three Nos' to China: no new island building, no violent contests of allied sovereignty, and no restricting the freedom of movement in the global commons."[200] To deter Chinese expansionism, the United States needs to support the nations of the Indo-Pacific with weapons and further cooperation through a multinational organization.

If the Americans are going to succeed in deterring China from war, they will have to understand and counter the Chinese strategy for

---

200 Faulhaber, Andrew (a captain in the U.S. Army, who has served in an array of tactical and strategic positions in both the military and U.S. government. He graduated from the Virginia Military Institute in 2018 with a BA in International Studies, History, and Arabic. He also received his master's in international relations from Harvard University in 2021.). "Deterring China's Military Aggression: DoD Strategic Options for Securing the Western Pacific." *Yale Journal of International Affairs*. Published by the Yale Jackson School at New Haven, Connecticut. May 15, 2023.

the South China Sea. The Chinese strategy in the region is covered by Wendell B. Leimbach and Eric Duckworth, who explain that "Chinese strategists call China's challenge to international norms 'war without gun smoke.'" The Chinese government is looking at alternatives to lethal warfare. One part of the 'war without gun smoke' strategy is the "coordinated use of media/public opinion; psychological warfare; and legal warfare to 'shape the battlespace by creating a favorable strategic and operational environment prior to hostilities.'" Leimbach and Duckworth suggest that legal warfare (sometimes called "lawfare") could be used to justify an attack on the enemy, while public opinion warfare delegitimizes the enemy and psychological warfare demoralizes them.[201] China has used these forces to support its aggressive actions without going to war. America is aware of these tactics, and could find a way to counter them.

The United States could use its ties with Australia and ASEAN to bolster a deterrence to China and encourage the Chinese to negotiate their claims in the South China Sea. This alternative is covered by Melanie Burton and Renju Jose. They note that the 2024 ASEAN-Australia Special Summit resulted in calls for restraint in the South China Sea and a lasting ceasefire in the Gaza Strip. Australia hosted a summit marking the 50th anniversary of its ties with ASEAN, despite the differences in the way participating nations respond to China. The summit resulted in a joint statement calling for a "rules-based" order in the Indo Pacific. The statement asserts, "We recognise the benefits of having the South China Sea as a sea of peace, stability, and prosperity … We encourage all countries to avoid any unilateral actions that endanger peace, security and stability in the region." Burton and Jose cite a spokesperson for the Chinese foreign ministry who said that "We will properly manage differences with the countries concerned

---

201 Leimbach, Wendell B. (Colonel in the U.S. Marine Corps), Duckworth, Eric (retired Lieutenant Colonel in the U.S. Army). "Prevailing without Gunsmoke in the South China Sea." *Proceedings* Vol. 148/11/1,437. Published by the U.S. Naval Institute in Annapolis, Maryland. November 2022.

and fully and effectively implement them with ASEAN countries."[202] China's talks with ASEAN could be a gateway for other nations to negotiate a territorial settlement.

As North Korea grows more independent of China, thanks to Russia, the concern over North Korea could be a unifying matter for America and China. Christy Lee claims that China could be facing a dilemma as North Korea is growing closer to Russia, and Kim Jong Un's nuclear threats could disrupt China's attempt to improve relations with America. She adds that 2024 saw North Korea refusing the possibility of reunification with South Korea and calling to occupy the south in the time of war. Afterwards, they conducted a test of an underwater nuclear weapons system. Meanwhile, North Korean foreign minister Choe Son-hui visited Russian President Vladimir Putin and Foreign Minister Sergey Lavrov to deepen ties between the two countries. Lee cites Dennis Wilder, the US National Security Council director for China from 2004 to 2005, who told VOA by email that China might be "nervous about its loss of influence over North Korea as it moves closer to Russia and 'worry about what kind of assistance in the military sphere Putin has promised Kim.'" North Korea is able to get away with saber-rattling toward the United States because China allows it to happen. China would not engage in a war at North Korea's behest, unless China is ready for war.[203] The gap between the relationships of China, Russia and North Korea might be wide enough for America to take advantage and could make China negotiate on how to deal with North Korea.

Despite China's aggression in the South China Sea, the United States government is determined to find a way to make peace with

---

202 Burton, Melanie (Asia Mining Correspondent at Thomson Reuters), Jose, Renju (Breaking News Correspondent, Reuters). "Australia and ASEAN call for restraint in South China Sea, ceasefire in Gaza." Reuters. March 6, 2024.

203 Lee, Christy (journalist who writes for Patheos, Inc. and VOA now Stantec). "North Korea's Hostility Could Snag China's Bid for Better US Ties, Analysts Say." Voice of America. January 19, 2024. https://www.voanews.com/a/north-korea-s-hostility-could-snag-china-s-bid-for-better-us-ties-analysts-say/7447992.html.

China and establish a future where both major world powers can engage in trade and diplomacy in the region. David Rising starts by citing former American Secretary of Defense Lloyd Austin, who said at the Shangri-La defense forum in Singapore that "war with China was neither imminent nor unavoidable, despite rapidly escalating tensions in the Asia-Pacific region, stressing the importance of renewed dialogue between him and his Chinese counterpart in avoiding 'miscalculations and misunderstandings.'" This came a day after Austin met with Chinese Defense Minister Dong Jun, which was the first in-person meeting with top defense officials since contact between the two militaries broke down in 2022. Neither side has changed its views on Taiwan. Austin argues that war is not imminent, but we need reduce the chances for miscommunication, by connecting further with China.[204] America's dialogue with China is slow but can open to negotiations in the future.

With the Russo-Ukrainian War still being waged as of this writing, Russia's difficulties in that conflict have increased China's caution in respect to invading Taiwan. Russell Hsiao and John Dotson claim that the Russo-Ukrainian War impacted China's calculation of success in an invasion in Taiwan, as the Chinese weigh not only America's response, but the response of Europe. Experts agree that a war between China and America would be costly on both sides, both militarily and financially. They also claim that a Ukrainian victory would give confidence to both Taiwan and America in terms of fighting a war against China. A Ukrainian victory would allow America to focus more attention on the South China Sea in order to contain Chinese aggression. A Russian victory, however, could see the American people

---

204 Rising, David (covers regional Asia-Pacific stories for The Associated Press. He has worked around the world, including covering the wars in Iraq, Afghanistan and Ukraine, and was based for nearly 20 years in Berlin before moving to Bangkok.). "US defense secretary says war with China neither imminent nor unavoidable, stressing need for talks." Associated Press. June 1, 2024. https://apnews.com/article/shangrila-dialogue-austin-china-taiwan-philippines-defense-46868bec3799f5043276d9fdeca62a41?utm_source=RecoReel&utm_medium=articlePage&utm_id=Taboola.

dissatisfied with aiding a far-off nation, and could lead the American military to reinforce Europe to counter an emboldened Russia.[205] A stalemate would be preferable in many ways. As Russia becomes mired in Ukraine, America could focus its efforts on Taiwan.

While every major power in the South China Sea is preparing for an eventual conflict, the buildup of military weapons and supplies has ironically made war in the region unappealing. It also helps that a naval war is harder to conduct than a land war. Not only that, but the economic consequences of war dissuade the nations of the region from engaging in open military conflict. The Russo-Ukrainian war has been a reminder to all nations of the cost of a major war involving one of the biggest nations in the world. The nations of the South China Sea need to come together for negotiations on overlapping claims and military exercises in order to prevent the breakdown of relations like what happened to Russia and Ukraine.

205 Hsiao and Dotson. 9, 11.

Chapter 17: Conclusion: A Cautiously Optimistic Look at the Future

Since the end of the Second World War, the situation in the South China Sea has been precarious at best. On one hand, the region has enjoyed an era of constant economic growth and with the surplus of resources in the South China Sea, the nations of said region will only grow further in the future. On the other hand, the nations have gone to war with each other for control of their pieces of the South China Sea and have armed themselves to back up their claims. China has taken a special interest in protecting its claims over Taiwan and the rest of the South China Sea by bullying its neighbors and making deals that keep them in debt. The United States has strengthened relations with the nations in the region and refocused its army toward the Indo-Pacific. This has created an air of tension and concern that is encompassing the world.

It is not helped by the zones of conflict that would be prevalent in a region-wide war. The Taiwan front is the most talked about area when it comes to potentially sparking the powder keg in the South China Sea. Taiwan is important to the US and its allies not only as a supplier of semiconductors, but also because of its location as the first line of defense against a Chinese invasion. For China, Taiwan remains an issue, as the government sees the island as a separatist province, which could inspire other groups like the Tibetans and the Uyghurs to similarly separate from China. There is also a fear that the Chinese people could look to Taiwan's ways of governance and call for reform in China, as the Chinese economy has stagnated in the first years of the 2020s and Chinese youth have grown dissatisfied with the Beijing government.

Another potential front in the South China Sea War is the Indo-Pakistani border, where both countries have held bitter relations with each other since the late 1940s. Not only do America and China support India and Pakistan respectively, but India's proximity to Tibet could spark a war on two fronts for China, which would distract them from the focus of Taiwan. The South China Sea War could also reignite the Korean War as the ties both Koreas have to America and China would make it difficult to avoid conflict in their peninsula. China's primary focus is Taiwan, where they would be fighting a naval war against the combined navies of America, Japan, Australia, and India. Even on the smallest scale, a South China Sea War would cause billions of dollars in damages and disrupt trade ties that help untold millions to live their lives.

However, as the past has shown us, continuing the peace in the South China Sea is a feasible objective. China has the most responsibility to maintain the free trade that allowed it to become one of the top economies of the world. The leaders of China need to realize the damage their nine-dash-lane plan has done to their country diplomatically as more of their neighbors prepare for war and are looking toward the nations of QUAD for help. The Chinese government should negotiate with whom they have overlapping claims to land and resources in the South China Sea. This would require China to cease their skirmishes with nations like the Philippines. Given their ties with India and Pakistan, China could settle its own claims in northwestern India, which could give China and Pakistan more resources to help the CPEC project. The Chinese government has the diplomatic resources to get all nations of the South China Sea together to discuss their claims; they need to be willing.

While Taiwan is in a dire situation with China, they have the tools to save themselves from going to war with their stronger neighbor. The

Taiwanese military can continue to take inspiration from Ukraine's defense against Russia, utilizing mobile weaponry such as javelin missiles and drones to create a better deterrence from China. As they watch their Russian allies' failure to achieve an efficient and speedy victory, the Chinese government will be wary of going into a war with an enemy employing similar tactics and weapons as Ukraine. Taiwan should persuade the QUAD to negotiate with China to reach a compromise on the Taiwan question. Taiwan could convince ASEAN to help in its defense, as a China-instigated conflict would not stop at Taiwan. Taiwan needs to defend itself while getting its partners to talk with China and try to end the tension between the two nations.

As America has economic and diplomatic ties to the South China Sea, it must continue to navigate through the political situation as it has done since the Second World War. The American government must strengthen their relations with allies in the QUAD and the nations of ASEAN to create a stronger deterrence against belligerence and to protect free trade and navigation. American representatives should meet with their Chinese and Russian equivalents to calm the tension among the three major powers in the South China Sea. Said meeting would have a positive impact on issues such as the Taiwan question; the coast guard clashes between China and the Philippines; the overlapping claims over the Kuril Islands between Russia and Japan; the denuclearization of North Korea; and the border clashes between China, Pakistan, and India. Although America should focus on making peace with China, it also needs to be prepared to defend itself, deter violence, and bring China to the negotiating table.

Though Japan is building up its defenses in case of a war with China, it must also find a way to maintain peace within the South China Sea. As a nation that has one of the longest-running relationships with China, Japan can connect to China on a deeper level. Japan could remind China of the thousands of years of trade relationships between the two cultures, and remind the Chinese how important this commerce is for both nations, historically and today. Japan would likely have to apologize for their actions against China during the Second Sino-Japanese War. The Japanese government could strengthen relations with South Korea through similar means, as Japan has an extensive history with Korea. As Japan builds up its navy, more exercises with ASEAN would present a stronger defensive front to China and could help deepen ties between QUAD and ASEAN. Because Japan is so close to China, it has the most to gain through keeping the peace in the South China Sea, but also the most to lose in the scenario of war.

South Korea, which is very close to the potential battleground, has a vested interest in keeping the waters calm in the South China Sea. Given their economic and military ties with China and the United States, South Korea could play a mediating role between the two nations and encourage them to settle their grievances with each other. South Korea has the capacity to encourage China and America to cooperate on their handling of North Korea. Meanwhile, the South Korean government needs to coordinate with the US and their partners to defend themselves from a war that could ignite on the South China Sea and on the Korean Peninsula. This would mean that South Korea and Japan would have to reconcile their relations, after decades of political spats. South Korea's neutrality gives them a plethora of opportunities that can help them and the region to escape the horrors of war.

India is one of the biggest players in the Indo-Pacific. Therefore, it has the potential to create a large impact for the region. Because of its border with China, India has a more direct way to connect to China than most nations in the South China Sea. This means that India could make diplomacy with China more effective, as a war would see the two countries occupying territory. India is geographically near to Tibet and then Xinjiang, which means their discussions with China would likely focus on those two regions. That is significant, given the number of Tibetans living in India and the international outcry against the Chinese government's treatment of Tibetans and Uyghurs. India has to contend with both China and Pakistan, as both nations strengthen their ties and have overlapping territorial claims with India. The Indian government could get help from America to mediate their discussions with Pakistan and China. India can be a great help in the South China Sea in terms of mending ties with China and having good relationships with America and its partners.

Given its mineral rich land and proximity to American and Chinese spheres of influence in the South China Sea, Australia has ways to leverage its resources to calm the region. Negotiations mediated by Australia could get it a better deal on mineral distribution with China and encourage America to focus more resources on the South China Sea, in return for Australia's support to both nations. Australia could deepen ties with ASEAN to protect itself and the organization in a scenario of war. Australia's partnership with the UK could lead to future support from European nations, which also are concerned over China's actions in the South China Sea. Australia can aid in maintaining the peace in the South China Sea by courting as many countries as they can.

While Russia is in a precarious place with the war in Ukraine and being isolated by Western tariffs, it can help itself and its neighbors in the South China Sea through diplomacy. The first thing it must do is to make peace with Ukraine and reconcile relations with the Western World. Russia especially must rebuild relations with the US if it wishes to have peace in the South China Sea. This would not only help Russia get out of a worsening situation but could assist its partners in talks over overlapping territorial claims and the denuclearization of North Korea. Given their connections with India, China and North Korea, Russian leaders could help mediate among their respective partners for deals that could benefit them all. Although Russia is in a rough spot, they could climb out of their rut and make a new path for cooperation among their partners and the other nations of the South China Sea.

While North Korea has gained the title of "Pariah State," its belligerence and nuclear weapons threats could provide a pathway for cooperation by China and the United States. The Americans and Chinese will have to negotiate with the North Koreans to denuclearize the Korean Peninsula in exchange for security guarantees and the reopening of trade with more nations. South Korea can also help in this matter as they have connections with China and America and would enjoy a warmer relationship with their neighbor. The issue with North Korea is being able to reconcile relations with them and break the status quo that North Korea has had for more than 70 years. However, all nations near them must try to reconnect with North Korea and make the Korean Peninsula a safer place to live.

As Pakistan stands to gain from economic cooperation, they would be at a loss if they were at war. The Pakistani government will need to focus on negotiating with their Indian counterpart on their claims over the Kashmir region. This could have India and Pakistan collaborate on counterterrorism efforts in the region. This would benefit China in terms of making things safer for their workers in Pakistan. It could also expand CPEC into India, creating greater interconnectedness among China, Pakistan, and India. Through their relationships with China, India and America, Pakistan could bring the three strongest nations in the South China Sea to the talking table. Stronger connections among the four countries would benefit Pakistan's ability to build up the economy and help its people. Although Pakistan has its own internal problems to deal with, they could help to keep the peace in the region.

ASEAN, as a neutral party in the South China Sea, would be one of the top groups that want peace in the region. ASEAN members are located throughout the region, which means that all Indo-Pacific trade must flow through the association's territory. This gives the organization an opportunity to mediate among the major powers. The nations of ASEAN need to court the QUAD nations and China to form a settlement on the issues of territorial claims and an agreement on how navies should conduct in the South China Sea. The nations of the organization should arm themselves to prepare for the possibility of war, as China has clashed with them in the past in the form of Coast Guard Skirmishes. As the Strait of Malacca could be a potential target in the South China Sea War, ASEAN has an interest in defending their most valuable asset. As ASEAN stands in the middle of the

Indo-Pacific, it can stand as a middle ground that the major powers of the region can utilize to discuss and create a safer Indo-Pacific.

European countries stand to gain through economic prosperity and peace as they look toward the Indo-Pacific for future economic prospects, and as China wishes to expand its ties with the continent. Europe will need to keep strengthening its relationships with nations in Southeast Asia to help defend their holdings in the South China Sea. The EU can use China's economic ambitions as leverage to bring China to the negotiating table and maintain the peaceful relations between themselves and China. The European nations could cooperate with the QUAD to maintain the freedom of trade and navigation that would benefit all parties involved. European powers could keep Russia occupied in a war with Ukraine to show their determination to help defend one of their neighbors. While Europe would take a more supportive role in the South China Sea, they would influence the region and help to maintain peace and free trade in the region.

Thanks to the relative peace the South China Sea has seen since the end of the Cold War, the leaders of the nations that make up the region have a path toward avoiding a war. It is important for all parties to be open to negotiation on their overlapping claims to land and resources in the South China Sea. The leaders must be aware that peace and free trade have been key elements in the economic growth their countries have had for decades. Any aggression would disrupt said peace and trade. America needs to come to terms with China, as past administrations were able to with the Soviet Union during the Cold War. For peace in the South China Sea to be maintained, China and America will have to cooperate in the spheres of trade, diplomacy, and the denuclearization of North Korea. This would provide the opportunity for the QUAD to work with China to ease tensions between each other and China's partners. The need for peace in the South China Sea would have to be backed up by the willpower of the leaders of the region to negotiate and keep the free trade that has helped the nations for more than 70 years.

References

Chapter One

Ahmad, Khadeeja (M.Phil. Scholar at the Department of Political Science, Government College University, Lahore, Pakistan), Rafique, Sadia (Assistant Professor at the Department of Political Science, Government College University, Lahore, Pakistan), Butt, Khalid (Director, Centre of Excellence China Studies, Chairperson, Department of Political Science and Dean, Faculty of Arts and Social Sciences at Government College University, Lahore, Pakistan.). "China's Policy towards South China Sea and its Geo-Political Impact (2009-2020). : War on Terror Partnership and Growing/Mounting/ Increasing/Rising Militant Extremism in Pakistan" (researchgate. net), A *Research Journal of South Asian Studies* Vol. 36, No. 2, July – December 2021 Published by the Centre for South Asian Studies, University of the Punjab. Lahore, Pakistan. 2022. 426.

Bartholomew, Carolyn (Chairman of the US-China Economic and Security Review Commission), Cleveland, Robin (Vice Chairman of the US-China Economic and Security Review Commission). "2021 Report to Congress." (The Commission was created on October 30, 2000, by the Floyd D. Spence National Defense). :2021 Annual Report to Congress | US- CHINA | ECONOMIC and SECURITY REVIEW COMMISSION (uscc.gov), US-China Economic and Security Review Commission One Hundred Seventeenth Congress First Session

November 2021. Published by US-China Economic and Security Review Commission. Washington, D.C., USA. November 2021.

Carter, Ash (a late American government official and academic who served as the 25th United States Secretary of Defense from February 2015 to January 2017), "Reflections on American Grand Strategy in Asia.": ReflectionsonAmericanGrandStrategyinAsia_0.pdf (belfercenter.org), Belfer Center Report. Published by Harvard Kennedy School Belfer Center for Science and International Affairs. Cambridge, Massachusetts, U.S.A. October 2018. 17.

Chen, Lung-chu (Scholar and Professor of Law at New York Law School, specializing in international law, human rights, the United Nations, and Taiwan). *The US-Taiwan-China Relationship in International Law and Policy.*: The US-Taiwan-China Relationship in International Law and Policy - Lung-chu Chen - Google Books. Published by Oxford University Press. New York, New York, U.S.A. 2016. XX.

Deepak, B. R. (Professor B. R. Deepak was trained in Chinese history and India-China relations at the Peking University and Chinese Academy of Social Sciences, Beijing, at Jawaharlal Nehru University, New Delhi and University of Edinburgh, U.K.). *China's Global Rebalancing and the New Silk Road*: Untitled-1 (springer.com), Published by Springer. New York, New York, U.S.A. January 22, 2018.

Dittmer, Lowell (Professor Dittmer received his Ph.D. from The University of Chicago in 1971. His scholarly expertise is the study of contemporary China. He teaches courses on contemporary China, Northeast Asia, and the Pacific Rim.). *Taiwan and China Fitful Embrace.* Published by the University of California Press (luminosoa.org) Oakland, California, U.S.A. 2017.

Dyer, Geoff A. (Geoff Dyer is an English author. He has written a number of novels and non-fiction books, some of which have won literary awards.). *The Contest of the Century: The New Era of Competition with China—and How America Can Win.*: The Contest of the Century: The New Era of Competition with China—and How … , Geoff A. Dyer, Google Books, Published by Vintage Books. New York, New York, U.S.A. 2014.

Ernst, Maximilian (a researcher with the Centre for Security, Diplomacy, and Strategy [CSDS] at the Brussels School of Governance and a PhD candidate at the Vrije Universiteit Brussels. His research focuses on Asia-Pacific security, particularly Chinese foreign policy towards regional states.). "Control Through Cooperation? Assessing China's Economic and Military-Strategic Interests in the South China Sea.": jtms_9.2.1.pdf (vub.be), *Journal of Territorial and Maritime Studies*. Published by Vrije Universiteit Brussel. Brussels, Belgium. 2022. 9-11.

Fanell, James E. (US Navy [Retired], served as a career naval intelligence officer whose positions included the senior intelligence officer for China at the Office of Naval Intelligence and the chief of intelligence for CTF-70, 7th Fleet, and the U.S. Pacific Fleet. He retired from the Navy in 2015 and currently is a government fellow at the Geneva Centre for Security Policy in Geneva, Switzerland.). "China's Global Naval Strategy and Expanding Force Structure: Pathway to Hegemony.": CHINA'S GLOBAL NAVAL STRATEGY AND EXPANDING FORCE STRUCTURE on JSTOR, *Naval War College Review*, Vol. 72, No. 1 (Winter 2019), Published by the US Naval War College. Newport, Rhode Island, U.S.A. 2019. 13-14, 40-41.

Gordon, Houlden (Professor Gordon Houlden is Director Emeritus of the China Institute, Professor of Political Science and Adjunct Professor of the Alberta School of Business at the University of

Alberta.), Romaniuk, Scott N. (Dr. Scott N. Romaniuk is a Newton International Fellow of the British Academy at the University of South Wales' Faculty of Life Sciences and Education, and a Visiting Fellow at the university's International Centre for Policing and Security), Hong, Nong (Executive Director of the Institute for China-America Studies and a Research Fellow at both the National Institute for South China Sea Studies, China, and the University of Alberta, Canada). *Security, Strategy, And Military Dynamics in the South China Sea: Cross-National Perspectives*. Published by Bristol University Press. Bristol, Britain. 2021. 297-298.

Jawli, Nandini (a freelance journalist, researcher, and a writer in Jaipur, India). "South China Sea and India's Geopolitical Interests.": South China Sea and India's Geopolitical Interests on JSTOR. *Indian Journal of Asian Affairs*, Vol. 29, No. 1/2 (June-December 2016). Published by the Indian Journal of Asian Affairs. Jaipur, India. 2016. 85-87, 89.

Kato, Yoshikazu (an adjunct associate professor at the Asia Global Institute, University of Hong Kong. He is also a research fellow at the Charhar Institute, and columnist for *The New York Times* [Chinese edition].). "US-Japan-China Trilateral Relations: How their dynamics will shape Asia Pacific.": asiaglobal-papers-us-china-japan-relations.pdf (hku.hk). Asiaglobal Papers: Working Paper. Published by Asia Global Institute, The University of Hong Kong. June 2019. 4-5.

Kuok, Lynn (fellow at Brookings Institution's Center for East Asia Policy Studies, a visiting fellow at the Harvard Law School, and a senior visiting fellow at the Centre for International Law). China Undermines Rule of Law in South China Sea: How China's actions in the South China Sea undermine the rule of law (brookings.edu), Global China Assessing China's Growing Role in The World, November 2019, Published by the Brookings Institution in Washington, D.C. III-10.

Li, Jiacheng (an associate professor at Liaoning University's School of International Studies in Liaoning, China). "China's Indian Ocean Strategy: Developing China's Indian Ocean Strategy: Rationale and Prospects." *China Quarterly of International Strategic Studies* Vol. 03, No. 04, (2017). Published by World Scientific Publishing Co Pte Ltd. Singapore. 483-484.

Lukin, Artyom (Associate Professor at Far Eastern Federal University School of Regional and International Studies at Vladivostok, Russia). China's Rejection of the South China Sea Arbitration: "The Russia–China entente and its future." *International Politics* Vol. 58 363-380. Published by the Moynihan Institute for Global Affairs, Syracuse University. New York, U.S.A. June 13, 2020.

McLaughlin, Madeline (the National Security Strategy intern at ASP. She graduated from the University of Utah in 2019 with a Fine Arts degree and a political science minor.). "US Strategy in the South China Sea.": resrep26608.pdf (jstor.org). American Security Project (2020). Published by American Security Project in Washington, D.C., U.S.A. October 2020.

Nie, Wenjuan (Deputy Director of Institute of International Relations, China Foreign Affairs University Beijing, China). One Belt One Road or the South China Sea: Xi Jinping's Foreign Policy Dilemma: One Belt, One Road or the South China Sea? on JSTOR, *Contemporary Southeast Asia*, Vol. 38, No. 3 (December 2016). Published by ISEAS - Yusof Ishak Institute. Heng Mui Keng Terrace, Singapore. December 2016. 424-433.

Pedrozo, Raul (Pete) (USN [Ret.], is the Howard S. Levie Professor of the Law of Armed Conflict at the Stockton Center for International Law, US Naval War College. After his retirement from the US Navy, he was a Professor of International Law at the Naval War College and

Principal Deputy Staff Judge Advocate, US Indo-Pacific Command.). "Does the Revised US South China Sea Policy Go Far Enough?": "Does the Revised US South China Sea Policy Go Far Enough?" by Raul (Pete) Pedrozo (usnwc.edu). *International Law Studies* Vol.99 2022. Published by the Stockton Center for International Law in North Carolina, U.S.A. 2022. 74, 77.

Ragi, Sangit (Sangit Kumar Ragi is currently a professor and Head in the Department of Political Science, University of Delhi.), *Indian Studies Review* Journal of Centre for Study of Politics and Governance Delhi Vol. 1 Number 2. July-December 2020: Publication (cspgindia.com), Published by the University of Delhi. Delhi, India. December 2020.

Scott, David (A doctor presenter on Indo-Pacific geopolitics at the NATO Defense College in Rome and the Baltic Defense College in Tartu, Estonia.). "US Indo-Pacific Strategy: The Indo-Pacific in US Strategy: Responding to Power Shifts - Rising Powers in Global Governance" (kureselcalismalar.com), *Rising Powers Quarterly* Volume 3, Issue 2, 2018, 19-43, Published by Marmara University. Istanbul, Turkey. August 2018.

Singh, Leishangthem. "China's Strategy in the South China Sea: China's Strategy On The South China Sea Disputes" on JSTOR. *World Affairs: The Journal of International Issues*, Vol. 21, No. 1 (SPRING JANUARY-MARCH 2017). Published by Kapur Surya Foundation. India. 84-85, 88.

Smith, Jeff M. (Research Fellow for South Asia in the Asian Studies Center in Pittsburgh, Pennsylvania). "UNCLOS: China, India, and the United States Navigate an Unsettled Regime.": BG3608.pdf (heritage.org). Backgrounder No. 3608 | April 30, 2021, Asian Studies Center.

Published by the Heritage Foundation, Washington, D.C., U.S.A. April 30, 2021. 16, 18.

Talmon, Stefan (Dr. Stefan Talmon LL.M. M.A. holds the Chair of Public Law, International Law and European Law. Prior to his appointment in October 2011, he was Lecturer, Reader and most recently Professor of Public International Law at the University of Oxford and Fellow of St. Anne's College, Oxford (2003-2011).), Jia, Bing (Professor of International Law Tsinghua Law School Beijing, China). *The South China Sea Arbitration: A Chinese Perspective*, Published by Hart Publishing 2014. 1-11.

Wuthnow, Joel (Senior Research Fellow in the Center for the Study of Chinese Military Affairs, Institute for National Strategic Studies, at the National Defense University). *System Overload: Can China's Military Be Distracted in a War over Taiwan?* System Overload: Can China's Military Be Distracted in a War over Taiwan? (ndu.edu). Published by Center for the Study of Chinese Military Affairs Institute for National Strategic Studies National Defense University. Washington, D.C., U.S.A. June 2020. 7.

Chapter Two

Bandurski, David (the Co-Director of the China Media Project, a research program in partnership with the Journalism & Media Studies Center at the University of Hong Kong). "The 'lying flat' movement standing in the way of China's innovation drive." Brookings Institution. July 8, 2021.

Cannon, Brendon J. (Assistant Professor of International Security at Khalifa University, Abu Dhabi, UAE), Rossiter, Ash (Dr. Ash Rossiter has been assistant professor of international security at Khalifa University, Abu Dhabi, UAE.). Rising-Powers-Quarterly-Volume-3-Issue-2.pdf (kureselcalismalar.com). *Rising Powers Quarterly*: The "Indo-Pacific" – Regional Dynamics in the 21st Century's New Geopolitical Center of Gravity Vol. 3, Issue 2|August 2018. Published by the Marmara University, Istanbul, Turkey. 2018. 71-73, 75.

Chubb, Andrew (British Academy Postdoctoral Fellow in the Department of Politics, Philosophy, and Religion at Lancaster University). "PRC Assertiveness in the South China Sea: Measuring Continuity and Change, 1970–2015.": PRC Assertiveness in the South China Sea: Measuring Continuity and Change, 1970–2015 | International Security | MIT Press. International Security (2021) 45 (3). Published by MIT Press. Cambridge, Massachusetts. 2021. 83, 87-90, 98.

Clarke, Michael (Associate Professor at the National Security College, Crawford School of Public Policy, Australian National University [ANU] and Director of the ANU-Indiana University Pan-Asia Institute). January 2019, "China's Applications of the Three Warfares: China's Application of the 'Three Warfares' in the South China Sea and Xinjiang." (anu.edu.au), Orbis Published for the Foreign Policy Research Institute by Elsevier Ltd. Amsterdam, Netherlands. Spring 2019. 14, 16-17, 19.

Fanell, James E. (US Navy [Retired], served as a career naval intelligence officer whose positions included the senior intelligence officer for China at the Office of Naval Intelligence and the chief of intelligence for CTF-70, 7th Fleet, and the US Pacific Fleet. He retired from the Navy in 2015 and currently is a government fellow at the Geneva Centre for Security Policy in Geneva, Switzerland.). "China's Global Naval Strategy and Expanding Force Structure: Pathway to Hegemony.": CHINA'S GLOBAL NAVAL STRATEGY AND EXPANDING FORCE

STRUCTURE on JSTOR, *Naval War College Review*, Vol. 72, No. 1 (Winter 2019), pp. 10-55, Published by the US Naval War College. Newport, Rhode Island, U.S.A. 2019. 14-22, 24-31, 33, 35-38.

Fravel, M. Taylor. *Active Defense: China's Military Strategy Since 1949.*: Active Defense: China's Military Strategy Since 1949—M. Taylor Fravel, Google Books, Published by Princeton University Press. Princeton, New Jersey, U.S.A. 2019. 217-219, 223, 226.

Ghiselli, Andrea (Assistant Professor at the School of International Relations and Public Affairs at Fudan University [China]). *Protecting China's Interests Overseas: Securitization and Foreign Policy.*: Protecting China's Interests Overseas: Securitization and Foreign Policy—Andrea Ghiselli, Google Books. Published at Oxford University Press. Oxford, Britain. 2021. 49.

Hiebert, Murray (A senior fellow and deputy director of the Sumitro Chair for Southeast Asia Studies at CSIS), Nguyen, Phuong (A research associate with the Sumitro Chair for Southeast Asia Studies at CSIS), Poling, Gregory B. (director of the Asia Maritime Transparency Initiative and a fellow with the Sumitro Chair for the Southeast Asia Studies at CSIS at Washington, D.C.). "Examining The South China Sea Disputes.": Examining the South China Sea Disputes: Papers from the Fifth Annual CSIS South China Sea Conference (csis-website-prod.s3.amazonaws.com). Papers from the Fifth Annual CSIS South China Sea Conference: A Report of the CSIS Sumitro Chair for Southeast Asia Studies. Published by the Center for Strategic & International Studies, Washington, D.C., U.S.A. September 2015. 4.

Jash, Amrita (Research Fellow at the Centre for Land Warfare Studies, New Delhi, India). "China's military and civil fusion.": Chinas-Military-Civil-Fusion-Strategy-Building-a-Strong-Nation-with-a-Strong-Military.pdf (researchgate.net). *CLAWS Journal* Winter 2020,

Published by the Centre for Land Warfare Studies (CLAWS). New Delhi, India. December 2020. 42-47, 54.

Kuntic, Dario (Minister Counselor at Permanent Representation of the Republic of Croatia to the European Union). "The Ominous Triangle: China-Taiwan-the United States relationship.": The Ominous Triangle: China-Taiwan-the United States relationship (srce.hr). *Croatian International Relations Review* 21. Published in Ministry of Science and Education of the Republic of Croatia. Zagreb, Croatia. March 2015. 247-248, 252-253.

Li, Jiacheng (an associate professor at Liaoning University's School of International Studies in Liaoning, China). "China's Indian Ocean Strategy: Developing China's Indian Ocean Strategy: Rationale and Prospects." *China Quarterly of International Strategic Studies* Vol. 03, No. 04, pp. 481-497 (2017). Published by World Scientific Publishing Co Pte Ltd. Singapore. 489-491, 494.

Miller, Tom (managing editor of the China Economic Quarterly and writes about China and India at Gavekal Research, a global economic research service based in Hong Kong). *China's Asian Dream: Empire Building along the New Silk Road*: China's Asian Dream: Empire Building along the New Silk Road - Tom Miller - Google Books. Published by Zed Books Ltd. London, Britain. 2019. XIII-XIV, XVI, XVIII-XX, 15, 19.

Poindexter, Dennis F. (has been a staff member in the US House of Representatives and made information warfare assessments for Ballistic Missile Defense, the National Command Authority and the Defense Threat Reduction Agency). *The Chinese Information War: Espionage, Cyberwar, Communications Control and Related Threats to United States Interests* 2d ed., :The Chinese Information War: Espionage, Cyberwar, Communications Control and ... —Dennis F.

Poindexter, Google Books, Published by McFarland Jefferson, North Carolina, U.S.A. 2nd edition (June 7, 2018). 6, 11-12, 16-17.

Roy, Denny (Ph.D. in political science, University of Chicago, 1991, writes mainly on Northeast Asian international security issues, particularly those involving China). *Return of the Dragon: Rising China and Regional Security.*: Return of the Dragon: Rising China and Regional Security - Denny Roy - Google Books. Published by Columbia University Press New York, New York, U.S.A. 2013. 15-18.

Sandy, Jordan M. (M.A. School of Public and International Affairs at the University of Princeton in New Jersey). "Chinese Nationalism and the South China Sea.": Ohio LINK ETD: Sandy, Jordan M. Published by Wright State University. Fairborn, Ohio, U.S.A. 2020. 21-22, 31, 34.

Sen, Rabindra (Professor at the Department of International Relations, Jadavpur University, Kolkata, India). "China's South China Sea Strategy: Balancing Triumphalism and Accommodation." IJEAS: *The International Journal of East Asian Studies* Vol. 7 No. 1 (2018).: China's South China Sea Strategy: Balancing Triumphalism and Accommodation | WILAYAH: The International Journal of East Asian Studies. Published by the University of Malay. Kuala Lumpur, Malaysia. December 15, 2018. 27, 31, 36, 40.

Turcsányi, Richard Q. (an Associate Research Fellow at the Institute of International Relations Prague who specializes in Chinese foreign policy, China's power and China's relations with Central Europe). *Chinese Assertiveness in the South China Sea: Power Sources, Domestic Politics, and Reactive Foreign Policy*: 978-3-319-67648-7.pdf (springer.com). Published by Springer. New York, New York, USA. 2018. 3-4, 10, 12, 43-50.

Wirth, Edward (Research Fellow at the GIGA German Institute of Global and Area Studies and an Adjunct Research Fellow at the Griffith Asia Institute). "Emotions, international hierarchy, and the problem of solipsism in Sino-US South China Sea politics." *International Relations*, Volume 34 Issue 1, March 2020. : Emotions, international hierarchy, and the problem of solipsism in Sino-US South China Sea politics - Christian Wirth, 2020 (sagepub.com), Published by Inquires Journal. Boston, Massachusetts, U.S.A. September 15, 2019. 5, 25-26, 31, 34, 38.

Worrell, Bradley (Writer and editor for the *Colorado Arts and Sciences Magazine*). "Is today's Ukraine tomorrow's Taiwan?" *Colorado Arts and Sciences Magazine*. Published by the University of Colorado in Boulder, Colorado. February 10, 2023. https://www.colorado.edu/asmagazine/2023/02/10/todays-ukraine-tomorrows-taiwan.

Zayaruzny, Dima (bachelor's in economics and Chinese at the University of California, Berkeley), Hoa, Kiet (Editor at the *Berkeley Economic Review*). "Lying Flat: The Demise of the Chinese Workforce and Its Impact on the Chinese Economy." *Berkeley Economic Review*. November 14, 2023.

Zhang, Hongzhou (A Research Fellow with the China Programme, Institute of Defence and Strategic Studies in Singapore), Bateman, Sam (A Professional Research Fellow at the Australian National Centre for Ocean Resources and Security at New South Wales, Australia). "Fishing Militia, the Securitization of Fishery and the South China Sea Dispute.": Fishing Militia, the Securitization of Fishery and the South China Sea Dispute on JSTOR. *Contemporary Southeast Asia* Vol. 39, No. 2 (August 2017), pp. 288-314. Published by ISEAS—Yusof Ishak Institute, Singapore. August 2017. 289, 292-293, 299.

Chapter Three

Bush, Richard C. (American expert on China affairs. Since 2002, he has served as the director of Center for Northeast Asian Policy Studies [CNAPS] of the Brookings Institution, and concurrently as the inaugural Michael H. Armacost Chair in Foreign Policy Studies.). *Difficult Choices: Taiwan's Quest for Security and the Good Life.*: Difficult Choices: Taiwan's Quest for Security and the Good Life - Richard C. Bush - Google Books. Published by the Brookings Institution Press. Wahington D.C., U.S.A. 2021. 126-129.

Carter, George (Author). "Projecting Taiwan's Military Manpower Supply, 2020-2065.": Title page for etd-1005121-004121 (nsysu.edu.tw). Global Human Resource Management English MBA Program. Published by National Sun Yat-sen University. Kaohsiung, Taiwan. November 2021. 6, 12, 22-23, 36-37, 85-88, 91-92, 100.

Chen, Lung-chu (Scholar and Professor of Law at New York Law School, specializing in international law, human rights, the United Nations, and Taiwan). *The US-Taiwan-China Relationship in International Law and Policy.*: The US-Taiwan-China Relationship in International Law and Policy - Lung-chu Chen, Google Books. Published by Oxford University Press. New York, New York, USA. 2016. 33.

Dotson, John (deputy director of the Global Taiwan Institute and associate editor of the *Global Taiwan Brief*). "Taiwan's 'Military Force Restructuring Plan' and the Extension of Conscripted Military Service.": https://globaltaiwan.org/2023/02/taiwan-military-force-restructuring-plan-and-the-extension-of-conscripted-military-service/. *Global Taiwan Brief* Vol 8. Issue 3. Published by the Global Taiwan Institute in Washington, D.C., U.S.A. February 8th, 2023. 2-3.

Greene, Andrew (ABC's defense correspondent and has covered

federal parliament since 2004), Grant, Stan (Vice Chancellor's Chair of Australian/Indigenous Belonging at Charles Sturt University. He was formerly ABC's Global Affairs and Indigenous Affairs Analyst.). "Taiwanese Foreign Minister warns his country is preparing for war with China, asks Australia for help." ABC. October 4, 2021. https://www.abc.net.au/news/2021-10-04/taiwan-preparing-for-war-with-china/100511294.

Grossman, Dereck (a senior defense analyst at the nonprofit, nonpartisan RAND Corporation. He formerly served as an intelligence adviser at the Pentagon.). "Ukraine War Is Motivating Taiwan to Better Secure Its Own Future." RAND Corporation. May 13, 2022. https://www.rand.org/pubs/commentary/2022/05/ukraine-war-is-motivating-taiwan-to-better-secure-its.html.

Hale, Erin (a journalist working at Al Jazeera). "Will Russia's struggle in Ukraine help Taiwan—or hurt it?" Al Jazeera. February 23, 2023. https://www.aljazeera.com/features/2023/2/23/will-russias-ukraine-struggle-help-taiwan-or-hurt.

Hsiao, Russell (the executive director of the Global Taiwan Institute [GTI] and editor-in-chief of the *Global Taiwan Brief*), Dotson, John (Associate Editor of the *Global Taiwan Brief*), Reid, Marshall (Staff Editor of the *Global Taiwan Brief*). *Global Taiwan Brief* Vol 8. Issue 13. Published by the Global Taiwan Institute in Washington, D.C. June 28, 2023. 1-2, 8, 13.

Roy, Denny (Senior Fellow and Supervisor of the POSCO Fellowship Program at the East-West Center in Honolulu). "Taiwan's Potential Role in The Free and Open Indo-Pacific Strategy Convergence in the South Pacific.": Taiwan's Potential Role in the Free and Open Indo-Pacific Strategy (nbr.org), NBR Special Report #77. Published by the National Bureau of Asian Research. Seattle Washington, U.S.A. March 2019.

Yang, William (an East Asia Correspondent, with his work appearing in various publications including VOA now Stantec). "Taiwan Faces Sovereignty Dilemma in South China Sea Amid Chinese Pressure." Voice of America. March 22, 2024.

Chapter Four

Hong, Nong (Executive Director Senior Fellow, Institute for China-America Studies). "Understanding the Freedom of Navigation Doctrine and China-US Relations in the South China Sea: Legal Concepts, Practice, and Policy Implication.": Understanding-FONOP-Doctrine-Report-FINAL.pdf (chinaus-icas.org). Published by Institute for China-America Studies in 2017. 19, 22-23.

Li, Jiacheng (an associate professor at Liaoning University's School of International Studies in Liaoning, China). "China's Indian Ocean Strategy: Developing China's Indian Ocean Strategy: Rationale and Prospects." *China Quarterly of International Strategic Studies* Vol. 03, No. 04, pp. 481-497 (2017). Published by World Scientific Publishing Co Pte Ltd. Singapore. 485-487.

Pomfret, John (American Journalist and Writer). *The Beautiful Country and the Middle Kingdom: America and China, 1776 to the Present.*: The Beautiful Country and the Middle Kingdom: America and China, 1776 to the ... - John Pomfret, Google Books. Published by Jenry Holt and Company, New York, New York, U.S.A. 2016. 2-3.

Simon, Sheldon W. (1937-2021) (Sheldon Weiss Simon, Professor Emeritus of Political Science at Arizona State University, who passed away on January 2, 2021). "The US Rebalance and Southeast Asia.": The US Rebalance and Southeast Asia on JSTOR. *Asian Survey* Vol.

55, No. 3 (May/June 2015), pp. 572-595. Published by University of California Press, Oakland, California, U.S.A. 2015. 474-475, 577, 579.

Zabelin, Dimitri (Associate at LSE IDEAS, having studied political economy at the University of California, Berkeley for both his undergraduate and masters). "How Can America Challenge China's Political Ambitions in an Age of Deglobalisation?": How can America challenge China's political ambitions in an age of deglobalisation? - *LSE Research Online*, LSE IDEAS Strategic Update September 2021. Published by LSE Ideas. London, Britain. September 2021. 6, 8-10.

Chapter Five

Burke, Edmund J. (senior international defense researcher at RAND. B.A. in government, Georgetown University), Heath, Timothy R. (Timothy R. Heath is a senior international defense researcher at the RAND Corporation. Prior to joining RAND, Heath had over fifteen years of experience in the US government researching and analyzing military and political topics related to China.), Hornung, Jeffery W. (senior political scientist at the RAND Corporation. He specializes in Japanese security and foreign policies, East Asian security issues, and US foreign and defense policies in the Indo-Pacific region, including its alliances. Prior to joining RAND in April 2017, Hornung was the fellow for the Security and Foreign Affairs Program at Sasakawa USA from 2015 until 2017.), Ma, Logan (adjunct research assistant at RAND. He is currently a master's student at the Johns Hopkins School of Advanced International Studies.), Morris, Lyle J. (Senior Fellow for Foreign Policy and National Security at Asia Society Policy Institute's Center for China Analysis. Prior to joining ASPI, Lyle was a senior policy analyst at the RAND Corporation leading projects on Chinese military modernization and Asia-Pacific security from 2011-2022.),

Chase, Michael S. (became the Deputy Assistant Secretary of Defense for China on 8 February 2021 and assumed the responsibilities for Taiwan and Mongolia in October 2022. He was previously a senior political scientist at RAND where his research focused on Chinese military modernization, China's approach to strategic deterrence, Taiwan's defense policy, and Indo-Pacific security issues). *China's Military Activities in the East China Sea: Implications for Japan's Air Self-Defense Force.*: China's Military Activities in the East China Sea: Implications for Japan's Air Self-Defense Force (rand.org). Published by the RAND Corporation. Santa Monica, California, U.S.A. 2018. VI, 1, 6-8.

Fatton, Lionel P., "'Japan is back': Autonomy and balancing amidst an unstable China–US–Japan triangle": "Japan is back": Autonomy and balancing amidst an unstable China–US–Japan triangle - Fatton - 2018 - Asia & the Pacific Policy Studies - Wiley Online Library, *Asia & the Pacific Policy Studies* Volume 5, Issue 2 Pages 165-386, Published by the Crawford School of Public Policy at the Australian National University in Canberra, Australia, May 2018. 264-266, 269-270, 273-274.

Miao, Ji (Associate Professor & Senior Research Fellow, Institute of Asian Studies, Center for China-Japan-Korea Trilateral Cooperation, China Foreign Affairs University). "Securing the Status Quo: Contesting Interests and Discourses in the East and South China Sea and Sino-Japan Relations.": 159504208.pdf (core.ac.uk). *Journal of Asia-Pacific Studies* (Waseda University) No. 30 (January 2018). Published by the institute of Asia-Pacific Studies at Waseda University. Tokyo, Japan. January 2018. 17-18, 22-23.

Midford, Paul (Paul Midford is an American political scientist who specializes in Japanese foreign policy. He teaches at the Meiji Gakuin University in Japan.). "Japan's Approach to Maritime Security in the South China Sea.": Japan's Approach to Maritime Security in the South

China Sea on JSTOR. *Asian Survey*, Vol. 55, No. 3 (May/June 2015), pp. 525-547. Published by University of California Press. Oakland, California, U.S.A. June 2015. 525-529, 535-540, 542-543.

Peng Er, Lam (political scientist, obtained his PhD from Columbia University. His publications have appeared in international journals such as the *Pacific Affairs, Asian Survey, Asian Affairs, Japan Forum and Government* and *Opposition: An International Journal of Comparative Politics.*). *China-Japan Relations in the 21st Century: Antagonism Despite Interdependency.*: China-Japan Relations in the 21st Century: Antagonism Despite Interdependency - Google Books. Published by Springer Nature. Singapore. 2017. 4-5, 7-9.

Rozman, Gilbert (an American sociologist specializing in Asian studies). Joint US-Korea Academic Studies: East Asian Leaders' Geopolitical Frameworks, New National Identity Impact, And Rising Economic Concerns with China 2020|Vol.31.: kei_jointus-korea_2020_final_digital_0.pdf (keia.org). Published by Korea Economic Institute of America, Washington, D.C., U.S.A. 2020. 66, 72, 76, 188, 191, 194, 210, 237, 243.

Shiraishi, Takashi (Chancellor, Prefectural University of Kumamoto; Professor Emeritus, National Graduate Institute for Policy Studies [GRIPS], Japan), Kojima, Takaaki (Visiting Senior Fellow at ISEAS and the former Japanese Ambassador to Singapore). *ASEAN-Japan Relations.*: ASEAN-Japan Relations - Google Books. Published by ISEAS. Singapore. 2014. II, 2-4, 17.

Smith, Sheila A. (John E. Merow senior fellow for Asia-Pacific studies at the Council on Foreign Relations [CFR]). *Japan Rearmed: The Politics of Military Power.*: Japan Rearmed: The Politics of Military Power - Sheila A. Smith - Google Books. Published by Harvard University Press. Cambridge, Massachusetts. 2019. 2, 5, 9, 12, 15.

Zulkifli, Noraini (works at the Department of International Relations, Security & Law, National Defense University of Malaysia.), Forbes, Vivian Louise (University of Western Australia · Australian Centre for Geomechanics [ACG] in Perth, Australia). "Japan and the South China Sea.": Japan and the South China Sea | Zulkifli | Andalas Journal of International Studies (AJIS) (unand.ac.id) *Andalas Journal of International Studies* Volume 5 NO.1 Mei Tahun 2016. Published by Andalas Institute of International Studies at Andalas University. Padang, Indonesia. May 2016. 63-64, 76-78.

Chapter Six

Aini, Resi (Lecturer and Researcher at the University of Indonesia), Triantama, Febry (Researcher and Lecturer with a demonstrated history of working both in the Public Policy Industry [Think Tanks] and University. Academia with a bachelor's degree in international relations from Paramadina University and a master's degree in international relations focusing on International Security from the University of Indonesia with The Indonesia Endowment Funds for Education [LPDP] Scholarships.). "The Implementation of South Korea's Military Technology Reform in The Perspective of Techno-nationalism.": The Implementation of South Korea's Military Technology Reform in The Perspective of Techno-nationalism | Sospol : Jurnal Sosial Politik (umm.ac.id), *Sospol Jurnal Sosial Politik* Vol. 7 No. 1 (2021): January-June. Published by the University of Muhammadiyah Malang. Malang, Indonesia. June 29, 2021. 64-66, 68, 71-73.

Ho, Chung Jae (a professor of international relations and director of the Institute for Chinese Studies at Seoul National University, was nominated by South Korean President Yoon Suk-yeol as the new ambassador to China on 2022). "South Korea's Strategic Approach to

China (or Lack of It)." Joint US-Korea Economic Studies.: jukas_1.4_south_koreas_strategic_approach_to_china.pdf (keia.org). Published by the Korean Economic Institute. Washington, D.C., U.S.A. August 13, 2018. 74-79, 81.

Hubbard, Corey J. (a researcher and author affiliated with the University of Navarra), Brito, Paula (bachelor's degree in international relations at the University of Navarra).: ROK_SAR_2021-April_CHubbard-PMora.pdf (unav.edu). *South Korea Strategic Analysis Report/April 2021: How the Inter-Korean Conflict is an indication of the New Cold War*. Published by the Center for Global Affairs & Strategic Studies at the University of Navarra. Pamplona, Spain. 2021. 19-21, 26, 28-31, 33-34, 38.

Lee, Seong-Hyon (senior fellow at the George H. W. Bush Foundation for US-China Relations and visiting scholar at the Harvard University Asia Center). "US-South Korea Coordination Toward China: Maritime Security." Council on Foreign Relations. January 4, 2024. https://www.cfr.org/blog/us-south-korea-coordination-toward-china-maritime-security.

Panneerselvam, Prakash (an Assistant Professor with the International Strategic & Security Studies Programme, National Institute of Advanced Studies [NIAS], Bengaluru, India). "Changing Dynamics of Regional Trade and Cooperation: A Case of Japan and South Korea.": Changing Dynamics of Regional Trade and Cooperation: A Case of Japan and South Korea | Artha Journal of Social Sciences (christuniversity.in). VOL. 19 NO. 4 (2020): *ARTHA Journal of Social Sciences*. Published by Christ University, Bangalore, India. 2020. 1-2, 4-6, 8-12.

Roehrig, Terrence (Ph.D. Professor National Security Affairs Biography Professor of national security affairs). "Caught in the Middle: South Korea and the South China Sea Arbitration Decision.": Caught in the

Middle: South Korea and the South China Sea Arbitration Decision in: Asian Yearbook of International Law, Volume 21 (2015) (brill.com), *Asian Yearbook of International Law*, Volume 21 (2015) 96-120. Published in Foundation for the Development of International Law in Asia (DILA). The Hague, Netherlands. September 17, 2017. 97-113.

Rozman, Gilbert (an American sociologist specializing in Asian studies). *Joint U.S.-Korea Academic Studies: East Asian Leaders' Geopolitical Frameworks, New National Identity Impact, And Rising Economic Concerns with China* 2020|Vol.31.: kei_jointus-korea_2020_final_digital_0.pdf (keia.org). Published by Korea Economic Institute of America, Washington, D.C., U.S.A. 2020. 98-100, 105-107.

Shin, Hyonshee (Journalist for Reuters). "South Korea fires warning shots after North's boat crosses sea border." Reuters. https://www.reuters.com/world/asia-pacific/south-korea-fires-warning-shots-after-norths-boat-crosses-sea-border-2023-04-16/.

Chapter Seven

Baruah, Darshana M. (fellow with the South Asia Program at the Carnegie Endowment for International Peace where she directs the Indian Ocean Initiative). "India's Answer to the Belt and Road: A Road Map for South Asia": resrep20723.pdf (jstor.org), Carnegie India Working Paper August 2018, Published by Carnegie India. New Delhi, India. 2018. 20-22.

Baruah, Darshana M. (a fellow with the South Asia Program at the Carnegie Endowment for International Peace where she directs the Indian Ocean Initiative). "India in the Indo-Pacific: New Delhi's Theater of Opportunity": Baruah_UnderstandingIndia_final1.pdf (carnegieendowment.org), Carnegie Endowment for International Peace Working Paper June 2020, Published by Carnegie Endowment for International Peace. Washington, D.C., U.S.A. 2020. 5-18.

Clary, Christopher (Christopher Clary is an Associate Professor of Political Science at the University of Albany.), Narang, Vipin (Vipin Narang is the Principal Deputy Assistant Secretary of Defense for Space Policy. Narang has been an American political scientist. He is Professor of Political Science at MIT.). "India's Counterforce Temptations: Strategic Dilemmas, Doctrine, and Capabilities": India's Counterforce Temptations: Strategic Dilemmas, Doctrine, and Capabilities | International Security | MIT Press. *International Security* (2019) Volume 43 Issue 3: 7–52. Published by MIT Press. Cambridge, Massachusetts, U.S.A. February 1, 2019. 7-9, 11-12, 15-16.

Fahmeed, Romana (Research Scholar, Department of Political Science, Bahauddin Zakariya University, Multan, Pakistan), Bukhari, Shahid (Associate Professor, Department of International Relations, Bahauddin Zakariya University, Multan, Pakistan), Ahmad, Shakeel (Research Fellow, Institute of Air and Space Law, McGill University, Canada). "India's Military Procurement Programs & Economic Capacity: Compatibility & Pragmatism.": India's Military Procurement Programs & Economic Capacity: Compatibility & Pragmatism | Journal of Business and Social Review in Emerging Economies (globalcsrc. org). *Journal of Business and Social Review in Emerging Economies* Vol. 6 No 1 (2020). Published by CSRC Publishing, Punjab, Pakistan. March 31, 2020. 135-136, 142.

Jaishankar, Dhruva (Executive Director of the Observer Research Foundation America [ORF America], which he helped establish in 2020. He is also a Non-Resident Fellow with the Lowy Institute in Australia and writes a regular column for the *Hindustan Times*.). "Survey of India's Strategic Community.": Survey of India's Strategic

Community (think-asia.org). Impact Series | March 2019. Published by Brookings Institution India Center. New Delhi, India. 2019. 3, 10, 13, 17, 21, 24.

Khattak, Masood (was a Pakistani civilian intelligence officer and the first Director General of the Intelligence Bureau [I.B.]). "Indian Military Modernisation: Implications for Pakistan.": Indian Military Modernisation on JSTOR. *Strategic Studies*, Vol. 39, No. 1 (Spring 2019), pp. 20-40. Published by International Institute for Strategic Studies. London, Britain. 2019. 21-32.

Mahajan, Utkarsha (Doctoral candidate at the Department of geopolitics and international relations, Manipal academy of higher education. Previously she has completed her Post graduation (M.A.) in Politics and Bachelor of Engineering in electronics and telecommunication from University of Mumbai along with two PG Diplomas in International and Security Studies.). India's Engagement in the Indo-pacific: A National Security Perspective.: INDIAS-ENGAGEMENT-IN-THE-INDO-PACIFIC-A-NATIONAL-SECURITY-PERSPECTIVE-Revised-as-of-2nd-Feb-2021.pdf (finsindia.org), Published by the Forum of integrated National Security (FINS). Dadar, India. April 28, 2021. 1, 5-8, 12-13.

Nachiappan, Karthik (a Fellow at the Institute of South Asian Studies, National University of Singapore). "India–ASEAN Relations: Riding and Transcending the 'Indo-Pacific' Wave.": India–ASEAN Relations: Riding and Transcending the "Indo-Pacific" Wave | The Journal of Indian and Asian Studies (worldscientific.com). *The Journal of Indian and Asian Studies* Vol. 02, No. 02, 2140004 (2021). Published by SAGE Publications, New York, New York, U.S.A. 2021. 2-3, 5-6, 10-12.

Nagao, Satoru (a fellow [non-resident] at Hudson Institute, based in Tokyo, Japan.). "Why Japan Needs India as a Defence Partner." India and Japan: Assessing the Strategic Partnership. Pp. 65-89. Published by Palgrave Pivot. Singapore. March 21, 2018. V, 2-3, 5-7, 14-15, 25-26, 31-32, 34-36, 55-58, 72, 76, 78-80, 92, 96-97, 99.

Sahoo, Prasanta (PhD Advanced International Relations Theory). "India's Balochistan Tactic: HAS IT SHATTERED PAKISTAN'S KASHMIR DREAM?": India's Balochistan Tactic on JSTOR. *World Affairs: The Journal of International Issues*, Vol. 23, No. 2 (SUMMER 2019 APRIL-JUNE), pp. 98-115. Published by World Affairs. New Delhi, India. June 2019. 98-102, 108-111.

Scott, David (a member of the Center for International Maritime Security). "India's China Challenge: Foreign Policy Dilemmas Post-Galwan and Post-Covid.": India's China Challenge: Foreign Policy Dilemmas Post-Galwan and Post-Covid | The Journal of Indian and Asian Studies (worldscientific.com). *The Journal of Indian and Asian Studies* Vol. 02, No. 02, (2021). Published by World Scientific Publishing Co Pte Ltd. Chennai, India. 2021. 5-11, 13, 16.

Shrikhande, Sudarshan (A retired Indian Admiral, a graduate of both the Soviet Naval War College (1988) and the US Naval War College (2003). He has served at sea and been defense attaché in Australia.). "Harnessing Indian Sea Power Post-Galwan: Considerations of Time, Space and Force.": ORF-Issue-rief-379-Sea-power.pdf (orfonline.org). *ORF Issue Brief*, July 2020, Issue No. 379. Published by Observer Research Foundation. New Delhi, India. July 10, 2020. 1-2, 4, 6, 9-10.

Singh, Abhijit (A former naval officer, Abhijit Singh, Senior Fellow, heads the Maritime Policy Initiative at ORF.). "The Nautical Dimension of India's 'Act East' Policy.": PR180409_The-Nautical-Dimension-of-Indias-Act-East-Policy.pdf (rsis.edu.sg), *Policy Report* April 2018. Published by S. Rajaratnam School of International Studies and Nanyang Technological School in Singapore, 2018. 4-6, 9.

Tarapore, Arzan (Fellow at NBR, adjunct at RAND. Former Australian Defence Department). "The Crisis after the Crisis: How Ladakh will Shape India's Competition with China.": TARAPORE Crisis after the Crisis PDF FINAL 05_05_2021(2).pdf (lowyinstitute.org).

Chapter Eight

Bell, Corey (Dr. Corey Lee Bell is a researcher at the Australia-China Relations Institute, University of Technology Sydney.), Collinson, Elena (head of analysis at the Australia-China Relations Institute, University of Technology Sydney). "Australia-China Relations: The Outlook for 2023." *The Diplomat*. December 22, 2022. https://thediplomat.com/2022/12/australia-china-relations-the-outlook-for-2023/.

Blaxland, John (professor and Head of the Strategic and Defence Studies Centre, in the Coral Bell School of Asia Pacific Affairs, College of Asia and the Pacific at the Australian National University). "Strategic Balancing Act: Australia's Approach to Managing China, the USA and Regional Security Priorities." *Security Challenges*, Vol. 13, No. 1. Published by The Institute for Regional Security in Barton, Australia. 19-20, 22-24, 29.

Cook, Dr. Malcolm (Nonresident Fellow at the Lowy Institute from 2010 to 2021. Completed a PhD in International Relations from the Research School of Pacific and Asian Studies at the Australian National University.). "Australia's South China Sea Challenges." Lowy Institute Policy Brief. Published by the Lowy Institute in Sydney, Australia. May 1, 2021. 2-4, 6, 13-14.

Crawley, Rhys (a senior lecturer in history at UNSW Canberra, and the author of the Official History of Australian Operations in Afghanistan,

2005-2010). "Australia's Lessons.": "Australia's Lessons" by Rhys Crawley (armywarcollege.edu). The US Army War College Quarterly: *Parameters* Winter 2019 Vol 49 Number 4 Article 8. Published by United States Army War College. Carlisle, Pennsylvania. November 1, 2019. 49, 52-53, 58, 60, 62.

Frost, Natasha (writes *The New York Times*'s weekday newsletter, "The Europe Morning Briefing," and reports on Australia, New Zealand and the Pacific from Melbourne, Also co-writes the "Australia Letter," a weekly newsletter about the region.). "Australia Tries to Break Its Dependence on China for Lithium Mining." *The New York Times*. May 23, 2023. https://www.nytimes.com/2023/05/23/business/australia-lithium-refining.html.

Glenn, Dr. Russell W. (a graduate of the United States Military Academy). New Directions in Strategic Thinking 2.0: ANU Strategic & Defense Studies Centre's Golden Anniversary Conference Proceedings.: New Directions in Strategic Thinking 2.0: ANU Strategic & Defense Studies Centre's Golden Anniversary Conference Proceedings (oapen.org), Published by Australian National University. Canberra, Australia. 2018. VIII, 30, 39, 43, 54.

Gomez, Jim (Associated Press Chief Correspondent based in Manila, Philippines), Aaron, Favila (Associated Press photographer). "US and Philippine forces sink a ship during largescale drills in the disputed South China Sea." Associated Press. May 9, 2024. https://apnews.com/article/philippines-china-us-military-drills-sea-ship-0157be2b96b90abad85b9c30b29e40a0.

Köllner, Patrick (vice president of the GIGA German Institute of Global and Area Studies, director of the GIGA Institute of Asian Studies, and professor of political science [with a focus on Asian studies] at the University of Hamburg). "Australia and New Zealand recalibrate their

China policies: convergence and divergence." *The Pacific Review* Vol. 34 issue 3 2021. Published by Routledge at Oxfordshire in Britain. 406. 411-412. 414-415.

McGuirk, Rod (Rod McGuirk is an Associated Press staff member based in Canberra, Australia.). "India, Australia commit to boosting strategic ties as their diplomats and defense chiefs hold talks." Associated Press. November 21, 2023. https://apnews.com/article/india-australia-talks-ccaecb76ae2d9e029aa80554b55f3e02.

Searight, Amy (a senior adviser and director of the Southeast Asia Program for The Center for Strategic and International Studies [CSIS]). "Countering China's Influence Activities Lessons from Australia." A Report of the CSIS Southeast Asia Program. Published by the CSIS in Washington, D.C. July 2020. 1-3, 5-9, 13, 25-27, 29-30.

Chapter Nine

Bolt, Paul J. (A professor of Political Science at the United States Air Force Academy in Colorado Springs), Cross, Sharyl N. (Director of the Kozmetsky Center at St. Edward's University in Austin and Global Policy scholar at the Kennan Institute Woodrow Wilson International Center for Scholars in Washington, D.C.). *China, Russia, and Twenty-First Century Global Geopolitics.*: China, Russia, and Twenty-First Century Global Geopolitics - Paul J. Bolt, Sharyl N. Cross - Google Books. Published by Oxford University Press. Oxford, Britain. 2018. 3, 7.

Bugos, Shannon (senior policy analyst at the Arms Control Association, where she conducts research, analysis, and media commentary on US-Russian nuclear arms control, nuclear disarmament and risk reduction, the US defense budget, China's nuclear arsenal, hypersonic weapons, and emerging technologies). "Russia: Russian Officials Talk Nuclear War, US Intelligence Says.": Russia | Arms Control Association. Arms Control Association. Published by Arms Control Association. Washington, D.C., U.S.A. December 2022.

Carlson, Brian G. (Former Head of the Global Security Team of the Think Tank at Center for Security Studies [CSS] at ETH Zürich). "China-Russia Relations and Asian Security.": China-Russia Relations and Asian Security - Research Collection (ethz.ch). Strategic Trends 2022: Key Developments in Global Affairs. Published by the Center of Security Studies. Zurich, Switzerland. May 13, 2022. 12-16, 18-19, 29-32.

Kirchberger, Sarah (the Head of Asia-Pacific Strategy & Security at ISPK. She also serves as Vice President of the German Maritime Institute [DMI] and is a Nonresident Senior Fellow at the Atlantic Council), Sinjen, Svenja (is head of science communication at the Foundation for Science and Democracy [SWuD] and an editor at the journal *SIRIUS*), Wörmer, Nils (is head of the Konrad-Adenauer-Stiftung's International and Security Affairs Department since September 2018). *Russia-China Relations Emerging Alliance or Eternal Rivals?* Published by Springer. New York City, New York. June 16, 2022. 3, 7, 17-20, 29, 35, 49-51, 54, 109-111, 114-116, 229, 236-238.

Muraviev, Alexey D. (Associate Professor of National Security and Strategic Studies at Curtin University, Perth, Western Australia). "Strategic reality check: the current state of Russia–China defence cooperation and the prospects of a deepening 'near alliance.'": Australian Journal of Defence and Strategic Studies Vol. 3 No.1 (curtin. edu.au). *Australian Journal of Defence and Strategic Studies* Volume 3, Number 1 (2021). Published by the Australian Defense Force. Canberra, Australia. July 1, 2021, 30-32, 35-36, 43.

Portyakov, Vladimir (is Professor and Deputy Director of the Institute of Far Eastern Affairs, Russian Academy of Sciences, Moscow, Russia). "Reluctant allies: system-unit dynamics and China-Russia relations.": Reluctant allies: system-unit dynamics and China-Russia relations - Alexander Korolev, Vladimir Portyakov, 2019 (sagepub.com). *International Relations* Volume 33, Issue 1, March 2019, Pages 40-66. Published by SAGE Publications. Thousand Oaks, California. March 2019. 43, 51, 53.

Rumer, Eugene (senior fellow and the director of Carnegie's Russia and Eurasia Program), Sokolsky, Richard (nonresident senior fellow in Carnegie's Russia and Eurasia Program, worked in the State Department for six different administrations), Vladicic, Aleksander (PhD student in the Department of Government at Cornell University). "Russia in the Asia-Pacific: Less Than Meets the Eye.": SokolskyRumer_Asia-Pacific_FINAL.pdf (carnegieendowment.org). The Return of Global Russia| September 2020. Published by Carnegie Endowment for International Peace. Washington, D.C., U.S.A. 2020. 4, 17, 20-21, 25-26, 30-31, 35.

Talukdar, Indrani (Research Fellow at the Indian Council of World Affairs, New Delhi). "Russia and China Nexus: Impact on India.": 201901-article3.pdf (cam.ac.uk). *Cambridge Journal of China Studies* Volume 14, No. 1 pg. 39-59. Published by the University of Cambridge, Cambridge, Britain. 2020. 46-47, 49-51.

Chapter Ten

Jones II, Stevie L. (Lieutenant, United States Navy). "China and North Korea: The Leverage Dilemma.": China and North Korea: The Leverage Dilemma (dtic.mil). Naval Postgraduate School: Thesis. Published by the Postgraduate Naval School. Monterey, California. March 1, 2018. 2-3, 13, 15-20, 23, 28-29, 36, 39-41, 50-51, 77.

McEachern, Patrick (Deputy Chief, Foreign Policy and Bilateral Affairs Unit, US Embassy Tokyo - US Embassy in Tokyo; Former North Korea Analyst, US State Department's Bureau of Intelligence and Research). *North Korea: What everyone needs to know.*: North Korea: What Everyone Needs to Know - Patrick McEachern - Google Books. Published by Oxford University Press. New York, New York. 2019. 102-104.

Shim, Kyu-Seok. "North Korea blasts Pompeo for South China Sea remarks." *Korea JoongAng Daily*. July 15, 2020. https://koreajoongangdaily.joins.com/2020/07/15/national/northKorea/North-Korea-China-South-China-Sea/20200715174000417.html.

Tasic, Mirko (lecturer and researcher at Webster University's Thailand campus in Bangkok, Thailand). "Exploring North Korea's Asymmetric Military Strategy." *Naval War College Review*, Vol. 72, No. 4 (Autumn 2019). Published by the US Naval War College in Newport, Rhode Island. 53-54, 56-61, 64.

Chapter Eleven

Allauddin (PhD Candidate at School of International Relations and Public Affairs, Shanghai International Studies University in Shanghai, China), Liu, Hongsong (Professor at School of International Relations and Public Affairs, Shanghai International Studies University), Ahmed, Raja Qaiser (Assistant Professor at School of Politics & IR, Quaid I Azam University Islamabad, Pakistan). "The Changing Dynamics and New Developments of China–Pakistan Relations.": The Changing

Dynamics and New Developments of China–Pakistan Relations - Allauddin, Hongsong Liu, Raja Qaiser Ahmed, 2020 (sagepub.com). *India Quarterly* Volume 76, Issue 1, March 2020, Pages 73-88. Published by SAGE Publications India. New Delhi, India. March 2020. 73-74, 76-77, 79, 83-86.

Fazl-E-Haider, Syed (a contributing analyst at the South Asia desk of Wikistrat. He is a freelance columnist and the author of several books, including *The Economic Development of Balochistan.*). "China's big gamble in Pakistan: A 10-year scorecard for CPEC." *The Interpreter*. August 1, 2023. https://www.lowyinstitute.org/the-interpreter/china-s-big-gamble-pakistan-10-year-scorecard-cpec.

Hassan, Kiran (University of London · School of Advanced Study). "CPEC: A win-win for China and Pakistan": CPEC: A win-win for China and Pakistan (degruyter.com). *Human Affairs* Volume 30 Issue 2. Published by De Gruyter. Berlin, Germany. April 15, 2020. 212-218, 221.

Hussain, Abid (Al Jazeera English's digital correspondent in Islamabad, Pakistan). "China to 'deepen and expand' military ties with Pakistan." Al Jazeera. April 27, 2023. https://www.aljazeera.com/news/2023/4/27/china-to-deepen-and-expand-military-ties-with-pakistan.

Iqbal, Saghir (a researcher in International Relations and Security Studies in the University of Lynchburg in Lynchburg, Virginia). *Pakistan's War Machine: An Encyclopedia of its Weapons, Strategy and Military Security* 2018 1st Edition.: Pakistan's War Machine: An Encyclopedia of its Weapons, Strategy and ... - Saghir Iqbal - Google Books. Published by CreateSpace Independent Publishing Platform. Scotts Valley, California, U.S.A. April 13, 2018. 5, 7, 12, 14, 63-64.

Miller, Manjari Chatterjee (senior fellow for India, Pakistan, and

South Asia at the Council on Foreign Relations [CFR]. She is also a research associate in the Contemporary South Asian Studies Programme at the Oxford School of Global and Area Studies at the University of Oxford.). "How China and Pakistan Forged Close Ties." *Hindustan Times*. October 3, 2022. https://www.cfr.org/article/how-china-and-pakistan-forged-close-ties.

Ragi, Sangit (Sangit Kumar Ragi is currently a professor and Head in the Department of Political Science, University of Delhi.). *Indian Studies Review* Journal of Centre for Study of Politics and Governance Delhi Vol. 1 Number 2. July-December 2020: Publication (cspgindia.com), Published by the University of Delhi. Delhi, India. December 2020. 85-87.

Chapter Twelve

Austin III, Lloyd J. "JOINT STATEMENT: Australia – Japan – Philippines – United States Maritime Cooperative Activity." April 4, 2024. https://www.defense.gov/News/Releases/Release/Article/3733095/joint-statement-australia-japan-philippines-united-states-maritime-cooperative/.

Bartholomew, Carolyn (Chairman of the US-China Economic and Security Review Commission), Cleveland, Robin (Vice Chairman of the US-China Economic and Security Review Commission). "2021 Report to Congress." (The Commission was created on October 30, 2000, by the Floyd D. Spence National Defense).: 2021 Annual Report to Congress | US- CHINA | ECONOMIC and SECURITY REVIEW COMMISSION (uscc.gov), US-China Economic and Security Review Commission One Hundred Seventeenth Congress First Session

November 2021. Published by US-China Economic and Security Review Commission. Washington, D.C., U.S.A. November 2021. 120.

Brands, Hal (American political scientist and writer. He is an international relations scholar of US foreign policy.). *American Grand Strategy in the Age of Trump.*: American Grand Strategy in the Age of Trump - Hal Brands - Google Books. Published by the Bookings Institution Press at Washington, D.C., U.S.A. 2018. 3, 14, 16.

Dziedzic, Stephen (is the ABC's Foreign Affairs [Asia Pacific] reporter, based in the Parliament House bureau. He covers foreign policy and Australia's relationship with countries in the Asia Pacific region. Stephen has worked for the ABC since 2007 and spent five years covering federal politics.). "Vladimir Putin's invasion of Ukraine neatly exposes the divisions and limitations of Quad nations." ABC News. March 4, 2022. https://www.abc.net.au/news/2022-03-04/quad-nations-divisions-exposed-by-putins-invasion-of-ukraine/100882258.

Hashimi, Sana (is Visiting Fellow at Taiwan-Asia Exchange Foundation. She is also an affiliated scholar at Research Institute for Indo-Pacific Affairs, Japan.). "Revival of the Quad: From Divergences to Convergences.": 743f685af69445c89c070adaecd002f1.pdf (mjib.gov.tw). *Prospect & Exploration* Volume 20, Issue 1 84-107. Published by Ministry of Justice Investigation Bureau. New Taipei, Taiwan. 84-87, 89-92.

Lyons, John (Global Affairs Editor of the ABC. John has previously been the Editor of *The Sydney Morning Herald*, Executive Producer of the Nine Network's Sunday program and Head of Current Affairs and Investigations at the ABC.). "Will Australia support the US in a war between Taiwan and China? Ask the military chiefs, not our political leaders." ABC. August 14, 2022. https://www.abc.net.au/news/2022-08-15/will-australia-join-the-us-in-a-war-between-taiwan-and-china-/101328658.

McGuirk, Rod (is an Associated Press staff member based in Canberra, Australia). "Australia won't promise to side with US in Taiwan conflict." Associated Press. March 20, 2023. https://apnews.com/article/australia-nuclear-powered-submarines-taiwan-208be03cc40dab2baaeb50843e3c8da7.

McLaughlin, Madeline (the National Security Strategy intern at ASP. She graduated from the University of Utah in 2019 with a Fine Arts degree and a political science minor.). "US Strategy in the South China Sea.": resrep26608.pdf (jstor.org). American Security Project (2020). Published by American Security Project in Washington, D.C., U.S.A. October 2020. 4-6.

Rej, Abhijnan (a Fellow with Observer Research Foundation's Strategic Studies Programme). "Reclaiming the Indo-Pacific: A Political-Military Strategy for Quad 2.0.": Untitled-1 (orfonline.org). ORF Occasional Paper 147 March 2018. Published by Observer Research Foundation. New Delhi, India. 2018. 8, 10, 14-17.

Sevastopulo, Demetri (Demetri served as Washington Bureau Chief from 2015 to 2021.). "Australia vows to help US defend Taiwan from Chinese attacks." *Financial Times*. November 13, 2021. https://www.ft.com/content/231df882-6667-4145-bc92-d1a54bccf333.

Wade, Geoff (historian who has studied Sino-Asian ties and Chinese influence efforts). "Australia-Japan relations." Parliament of Australia. https://www.aph.gov.au/About_Parliament/Parliamentary_Departments/Parliamentary_Library/pubs/BriefingBook47p/AustraliaJapanRelations.

Chapter Thirteen

Coyne, Dr. John (Senior Analyst for the Border Security Program for the ASPI), Sharpe, Ashleigh (ASPI intern), Hodgson, Dione (ASPI intern). "Mice that roar: patrol and coastal combatants in ASEAN.": Mice that roar: patrol and coastal combatants in ASEAN (apo.org.au). ASPI: Special Report | August 2018. Published by the Australian Strategic Policy Institute. Canberra, Australia. August 1, 2018. 5-6, 8-9, 12, 14, 21, 26, 29-30.

El, Monineath (Tax Economist at Ministry of Economy and Finance at Phnom Penh, Cambodia). "Cambodia-between-China-and-Japan-in-east-China-sea-dispute.": Microsoft Office 2000 (researchgate.net). Royal University of Phnom Penh Department of International Studies Foreign Policy II: Cambodia Focus. Published by Royal University of Phnom Penh in Cambodia. 2015. 3-5, 9-10, 15-17, 19-20.

Emmers, Ralf (Senior Lecturer in International Politics of East Asia and Co-Chair of the Centre for International Studies and Diplomacy [CISD] in London, Britain). "ASEAN's Search for Neutrality in the South China Sea.":04_Ralf Emmers_DOI.indd (snu.ac.kr). *Asian Journal of Peacebuilding* Vol. 2 No. 1 (2014): 61-77. Published by IPUS at Seoul National University. Seoul, South Korea. 2014. 62, 64, 67.

Grossman, Dereck (a senior defense analyst at the RAND Corporation). "Can Vietnam's Military Stand Up to China in the South China Sea?.": Can Vietnam's Military Stand Up to China in the South China Sea? on JSTOR. *Asia Policy*, Vol. 13, No. 1 (JANUARY 2018), pp. 113-134. Published by the NBR. Seattle, Washington, U.S.A. January 2018.

Hiep, Le Hong (Dr. Le Hong Hiep is a Senior Fellow at the Vietnam Studies Programme.). "Vietnam's Hedging Strategy against China since Normalization.": Vietnam's Hedging Strategy against China since

Normalization on JSTOR. *Contemporary Southeast Asia*, Vol. 35, No. 3 (December 2013), pp. 333-368. Published by ISEAS–Yusof Ishak Institute. Singapore. December 2013. 334, 345-346, 357, 359.

Hiep, Le Hong (Dr. Le Hong Hiep is a Senior Fellow at the Vietnam Studies Programme and the Regional Strategic and Political Studies Programme of the ISEAS – Yusof Ishak Institute.). "Vietnam's Alliance Politics in the South China Sea.": trends_in_sea_2015_6.pdf (iseas.edu.sg). *Trends in Southeast Asia* 2015 #06. Published by IEAS. Singapore. 2015. 2-5, 9.

Hoang, Phuong (National Intelligence University, Washington, D.C.). "Domestic Protests and Foreign Policy: An Examination of Anti-China Protests in Vietnam and Vietnamese Policy Towards China Regarding the South China Sea.": Domestic Protests and Foreign Policy: An Examination of Anti-China Protests in Vietnam and Vietnamese Policy Towards China Regarding the South China Sea - Phuong Hoang, 2019 (sagepub.com). *Journal of Asian Security and International Affairs* Volume 6, Issue 1. Published by SAGE publications. Thousand Oaks, California, U.S.A. March 26, 2019. 2, 7, 9, 13, 15, 18, 20.

Kreuzer, Peter (Senior Research Fellow at the Peace Research Institute Frankfurt in Germany). "A Comparison of Malaysian and Philippine Responses to China in the South China Sea.": Comparison of Malaysian and Philippine Responses to China in the South China Sea | The Chinese Journal of International Politics | Oxford Academic (oup.com). *The Chinese Journal of International Politics*, Volume 9, Issue 3, Autumn 2016, Pages 239–276. Published by Oxford University Press. Oxford, Britain. June 16, 2016. 239, 249-250, 265, 269-270.

Panarina, Daria S. (Researcher, Center for Southeast Asia, Australia and Oceania, Institute of Oriental Studies, Moscow, Russia). "Why South China Sea is Important for the Philippines: Economic Aspect.":

SEA2020n1(46)p356-369.pdf (ivran.ru). *Southeast Asia: Actual Problems of Development* 2020, Volume I, No. 1 (46). pp. 356-369. Published By Institute of Oriental Studies Russian Academy of Sciences. Moscow, Russia. January 2020. 357-358, 360, 362-363.

Patunru, Arianto A. (A fellow at the Arndt-Corden Department of Economics and Policy Engagement Coordinator), Pangetsu, Mari (Professor of International Economics at the Faculty of Economics in the University of Indonesia in Jakarta, Indonesia), Basri, M. Chatib (Senior Lecturer at the Department of Economics in the University of Indonesia in Jakarta, Indonesia). *Indonesia in the New World: Globalisation, Nationalism and Sovereignty.*: Indonesia in the New World: Globalisation, Nationalism and Sovereignty - Google Books. Published by IEAS Yusof Ishak Institute. Singapore. 2018. 4, 58.

Rosen, Mark E. (Mark Rosen is an SVP and General Counsel at CNA Corporation.)."Philippine Claims in the South China Sea: A Legal Analysis": Philippine Claims in the South China Sea: A Legal Analysis (cna.org). A CNA Occasional Paper. Published by CNA analysis and solutions. Arlington County, Virginia, U.S.A. August 2014. I-II, VII, 7-12, 20-22.

Saha, Premesha (a Fellow with ORF's Strategic Studies Programme). "India Calibrates its South China Sea Approach.": ORF_IssueBrief_477_India-SouthChinaSea_FinalForUpload.pdf (orfonline.org). Issue Brief No. 477 July 2021. Published by Observer Research Foundation. New Delhi, India. 2021. 14-17.

Sandy, Jordan M. (M.A. School of Public and International Affairs at the Wright State University in Dayton, Ohio). "Chinese Nationalism and the South China Sea." CORE Scholar. Ohio LINK ETD: Sandy, Jordan M. Published by Wright State University. Fairborn, Ohio, U.S.A. 2020. 45-46, 52-53, 55, 59.

Ting, Wai (Emeritus Professor at Hong Kong Baptist University, Hong Kong). "Increasing Economic Interdependence between China and ASEAN and Its Implications.": Conflicts-in-the-South-China-Sea-and-China-ASEAN-Economic-Interdependence-A-Challenge-to-Cooperation.pdf (researchgate.net). *Journal of ASEAN Studies*, Vol. 4, No. 2 (2016), pp. 127-141. Published by CBDS Bina Nusantara University. Kebon Jeruk, Indonesia. 2016. 127-128, 132-133, 139.

Turcsányi, Richard Q. (an Associate Research Fellow at the Institute of International Relations Prague who specializes in Chinese foreign policy, China's power and China's relations with Central Europe). *Chinese Assertiveness in the South China Sea: Power Sources, Domestic Politics, and Reactive Foreign Policy*: 978-3-319-67648-7.pdf (springer.com). Published by Springer. New York, New York, U.S.A. 2018. 44-51, 53.

Chapter Fourteen

Gomez, Jim (Associated Press Chief Correspondent based in Manila, Philippines). "EU eyes more naval visits in disputed South China Sea." Associated Press. March 15, 2023. https://apnews.com/article/european-union-navy-south-china-sea-643ee2f29e52321fa9efaf481006f13c.

Gordon, Houlden (Professor Gordon Houlden is Director Emeritus of the China Institute, Professor of Political Science and Adjunct Professor of the Alberta School of Business at the University of Alberta), Romaniuk, Scott N. (Dr. Scott N. Romaniuk is a Newton International Fellow of the British Academy at the University of South Wales's Faculty of Life Sciences and Education, and a Visiting Fellow at the university's International Centre for Policing and Security), Hong,

Nong (Executive Director of the Institute for China-America Studies and a Research Fellow at both the National Institute for South China Sea Studies, China, and the University of Alberta, Canada). *Security, Strategy, And Military Dynamics in the South China Sea: Cross-National Perspectives.*: Security, Strategy, and Military Dynamics in the South China Sea: Cross ... - Google Books, Published by Bristol University Press. Bristol, Britain. 2021. 308-312, 316, 319.

Horta, Loro (deputy head of mission at the Timor Leste Embassy in Beijing. He is a graduate of the Chinese National Defense University senior officers' course, the US Naval Post Graduate School, the American National Defense University and the S Rajaratnam School of International Studies.). "Europe and the South China Sea." *The Interpreter*. September 2, 2021. https://www.lowyinstitute.org/the-interpreter/europe-south-china-sea.

Manoj, Joshi (a Distinguished Fellow at the Observer Research Foundation, New Delhi). "China and Europe: Trade, Technology and Competition.": OP#194-new text (orfonline.org). ORF Occasional Paper #194 May 2019. Published by Observer Research Foundation, New Delhi, India. May 2019. 3, 6, 9, 11, 16, 19-20.

Morcos, Pierre (a visiting fellow with the Europe, Russia, and Eurasia Program at the Center for Strategic and International Studies [CSIS] in Washington, D.C.). "France: A bridge between Europe and the Indo-Pacific?" Center for Strategic and International Studies. April 1, 2021. https://www.csis.org/analysis/france-bridge-between-europe-and-indo-pacific.

Odgaard, Liselotte (A Senior Fellow at the Hudson Institute at Washington, D.C.). "European Engagement in the Indo-Pacific.": European Engagement in the Indo-Pacific on JSTOR. *Asia Policy* Vol. 14, No. 4 (OCTOBER

2019), pp. 129-160. Published by National Bureau of Asian Research, New Delhi, India. October 2019. 135-137, 153-154.

Scott, David (A doctor presenter on Indo-Pacific geopolitics at the NATO Defense College in Rome and the Baltic Defence College in Tartu, Estonia.). "US Indo-Pacific Strategy: The Indo-Pacific in US Strategy: Responding to Power Shifts - Rising Powers in Global Governance" (kureselcalismalar.com), *Rising Powers Quarterly* Volume 3, Issue 2, 2018, 19-43, Published by Marmara University. Istanbul, Turkey. August 2018. 34.

Vanoverbeke, Dimitri (Has a PhD from KU Leuven and lectures in Japanese Studies at the Department of Area Studies at Leuven), Suami, Takao (Professor of Law at Waseda University Law School and served as the President of the Japanese Association of EU Studies), Ueta, Takako (professor at the International Christian University in Japan and was a Ambassador, Deputy Chief of the Mission of Japan to the EU), Peeters, Nicholas (Studied Law at the University of Innsbruck and was the Minister/Deputy Head of the EU Delegation to Japan from 2002 to 2006), Ponjaert, Frederik (Researcher and Lecturer at the Institute of European Studies at the ULB and KU Leuven). *Developing EU–Japan Relations in a Changing Regional Context: A Focus on Security, Law, and Policies.*: Developing EU–Japan Relations in a Changing Regional Context: A Focus on ... - Google Books. The Globalisation, Europe and Multilateralism Series. Published by Routledge, New York, New York. 2018. XII, 2, 14-16, 22-23.

Walt, Stephen M. (a columnist at Foreign Policy and the Robert and Renée Belfer professor of international relations at Harvard University). "Will Europe Ever Really Confront China?" *Foreign Policy*. October 15, 2021. https://foreignpolicy.com/2021/10/15/will-europe-ever-really-confront-china/.

Chapter 15

Blackwill, Robert D. (the Henry A. Kissinger senior fellow for U.S. foreign policy at the Council on Foreign Relations). Zelikow, Philip (the White Burkett Miller professor of history and Wilson Newman professor of governance at the Miller Center, both at the University of Virginia). "The United States, China, and Taiwan: A Strategy to Prevent War." Council Special Report No. 90 February 2021. Published by the Council of Foreign Relations in New York, New York. 32-38.

Bowman, Dr. Gary (Geopolitical Conflict Project Lead), Caccioli, Dr. Fabio (Research Associate), Kelly, Dr. Scott (Research Associate), Foulser-Piggott, Dr. Roxane (Research Associate), Coburn, Dr. Andrew (Director of External Advisory Board, Professor), Ralph, Daniel (Academic Director), Ruffle, Simon (Director of Technology Research). "China-Japan Geopolitical Conflict Scenario": China-Japan Geopolitical Conflict Scenario (drscottkelly.com). Cambridge Centre for Risk Studies: Cambridge Risk Framework: Geopolitical Conflict: Stress Test Scenario. Published by the Cambridge Centre for Risk Studies, University of Cambridge Judge Business School. Cambridge, Britain. June 2014. 3, 38.

Clary, Christopher (Christopher Clary is an Associate Professor of Political Science at the University of Albany.), Narang, Vipin (Vipin Narang is the Principal Deputy Assistant Secretary of Defense for Space Policy. Narang has been an American political scientist. He is Professor of Political Science at MIT.). "India's Counterforce Temptations: Strategic Dilemmas, Doctrine, and Capabilities": India's Counterforce Temptations: Strategic Dilemmas, Doctrine, and Capabilities | International Security | MIT Press. *International Security* (2019) Volume 43 Issue 3: 7–52. Published by MIT Press. Cambridge, Massachusetts, U.S.A. February 1, 2019. 17-18, 21-22, 29-31, 39-40, 48.

Iqbal, Saghir (a researcher in International Relations and Security Studies in the University of Lynchburg in Lynchburg, Virginia). *Pakistan's War Machine: An Encyclopedia of its Weapons, Strategy and Military Security* 2018 1st Edition.: Pakistan's War Machine: An Encyclopedia of its Weapons, Strategy and ... - Saghir Iqbal - Google Books. Published by CreateSpace Independent Publishing Platform. Scotts Valley, California, U.S.A. April 13, 2018. 21-23, 25, 27-29.

Menski, Werner (Member of the SOAS South Asia Institute in London, Britain), Yousuf, Muneeb (a Kashmiri scholar who completed his master's in conflict analysis and peacebuilding from the Nelson Mandela Centre for Peace and Conflict Resolution, Jamia Millia Islamia University, New Delhi). *Kashmir after 2019: Completing the Partition.*: Kashmir after 2019: Completing the Partition - Google Books. Published by SAGE Publications India. New Delhi, India. 2022.

Sacks, David. "Why China Would Struggle to Invade Taiwan." Council on Foreign Relations. January 10, 2024. New York, New York. https://www.cfr.org/article/why-china-would-struggle-invade-taiwan.

Sway, Nikki (Senior at Bard College). "India's Military and its Impact on the Formation of a Nation.": India's Military and its Impact on the Formation of a Nation (bard.edu). Bard Master of Art in Teaching Program Academic Research Project. Published by Bard College. Annandale-On-Hudson, NY. January 28, 2020. 8.

Chapter Sixteen

Burton, Melanie (Asia Mining Correspondent at Thomson Reuters), Jose, Renju (Breaking News Correspondent, Reuters). "Australia and ASEAN call for restraint in South China Sea, ceasefire in Gaza." Reuters. March 6, 2024.

Faulhaber, Andrew (a captain in the US Army, who has served in an array of tactical and strategic positions in both the military and US government. He graduated from the Virginia Military Institute in 2018 with a BA in International Studies, History, and Arabic. He also received his master's in international relations from Harvard University in 2021.). "Deterring China's Military Aggression: DoD Strategic Options for Securing the Western Pacific." *Yale Journal of International Affairs*. Published by the Yale Jackson School at New Haven, Connecticut. May 15, 2023.

Garcia-Herrero, Alicia (Senior Fellow at Bruegel), Schindowski, Robin (A Research Assistant at Bruegel). "China's Quest for Innovation: Progress and Bottlenecks." Working Paper Issue 08/2023. Published by Bruegel in Brussels, Belgium. June 21, 2023. 1-2, 19-20, 21.

Gompert, David C., Binnendijk, Hans, Lin, Bonny (Members of the RAND Corporation). "Chapter Seventeen the Sino-US Case." *Blinders, Blunders, and Wars: What America and China Can Learn*. Published by RAND Corporation in Santa Monica, California. 2014, 217-218, 223, 225, 228-229, 239, 244-245.

Ikenberry, John G. (a theorist of international relations and United States foreign policy, and the Albert G. Milbank Professor of Politics and International Affairs at Princeton University). "Between the Eagle and the Dragon: America, China, and Middle State Strategies in East Asia.": Between the Eagle and the Dragon: America, China, and Middle State Strategies in East Asia on JSTOR. *Political Science Quarterly* Vol. 131, No. 1 (Spring 2016), pp. 9-43. Published by Oxford University Press. Oxford, Britain. 2016. 11.

Kaplan, Robert D. (bestselling author on foreign affairs and holds the Robert Strausz-Hupé Chair in Geopolitics at the Foreign Policy Research Institute. He was a member of the Pentagon's Defense Policy Board and the U. S. Navy's Executive Panel.). *Asia's Cauldron: The South China Sea and the End of a Stable Pacific* Asia's Cauldron: The South China Sea and the End of a Stable Pacific - Robert D. Kaplan - Google Books. Published by Random House LLC., Manhattan, New York. U.S.A. 2014. 7, 9-10.

Kennedy, Scott (senior adviser and Trustee Chair in Chinese Business and Economics at the Center for Strategic and International Studies [CSIS]. A leading authority on Chinese economic policy and US-China commercial relations, Kennedy has traveled to China for 36 years.). "US-China Relations in 2024: Managing Competition without Conflict." Center for Strategic and International Studies. January 3, 2024. https://www.csis.org/analysis/us-china-relations-2024-managing-competition-without-conflict.

Kirchberger, Sarah (the Head of Asia-Pacific Strategy & Security at ISPK. She also serves as Vice President of the German Maritime Institute [DMI] and is a Nonresident Senior Fellow at the Atlantic Council), Sinjen, Svenja (is head of science communication at the Foundation for Science and Democracy [SWuD] and an editor at the journal *SIRIUS*), Wörmer, Nils (is head of the Konrad-Adenauer-Stiftung's International and Security Affairs Department since September 2018). *Russia-China Relations Emerging Alliance or Eternal Rivals?* Published by Springer. New York City, New York. June 16, 2022. 279, 289.

Lee, Christy (journalist who writes for Patheos, Inc. and VOA now Stantec). "North Korea's Hostility Could Snag China's Bid for Better US Ties, Analysts Say." Voice of America. January 19, 2024. https://www.voanews.com/a/north-korea-s-hostility-could-snag-china-s-bid-for-better-us-ties-analysts-say/7447992.html.

Leimbach, Wendell B. (Colonel in the US Marine Corps), Duckworth, Eric (retired Lieutenant Colonel in the US Army). "Prevailing without Gunsmoke in the South China Sea." *Proceedings* Vol. 148/11/1,437. Published by the US Naval Institute in Annapolis, Maryland. November 2022.

Li, Jiacheng (an associate professor at Liaoning University's School of International Studies in Liaoning, China). "China's Indian Ocean Strategy: Developing China's Indian Ocean Strategy: Rationale and Prospects." *China Quarterly of International Strategic Studies* Vol. 03, No. 04, pp. 481-497 (2017). Published by World Scientific Publishing Co Pte Ltd. Singapore. 497.

Mastro, Oriana (Assistant Professor of Security Studies, Georgetown University's Edmund A. Walsh School of Foreign Service). "Military Confrontation in the South China Sea." Contingency Planning Memorandum No. 36. Published by the Council on Foreign Relations. May 21, 2020.

Rising, David (covers regional Asia-Pacific stories for The Associated Press. He has worked around the world, including covering the wars in Iraq, Afghanistan and Ukraine, and was based for nearly 20 years in Berlin before moving to Bangkok.). "US defense secretary says war with China neither imminent nor unavoidable, stressing need for talks." Associated Press. June 1, 2024. https://apnews.com/article/shangrila-dialogue-austin-china-taiwan-philippines-defense-46868bec3799f5043276d9fdeca62a41?utm_source=RecoReel&utm_medium=articlePage&utm_id=Taboola.

Strating, Rebecca (Senior Lecturer in Politics and International Relations in the Department of Politics, Media and Philosophy at La Trobe University, Melbourne, Australia). "Defending the Maritime Rules-Based Order: Regional Responses to the South China Sea

Disputes.": Defending the maritime rules-based order: regional responses to the South China Sea disputes (hawaii.edu). Policy Studies|80. Published by the East-West Center. Honolulu, Hawaii. 2020. 8, 13, 27-28.

Acknowledgments

I'd like to thank these people for their support:

- Kimberly Banker
- Tracey Banker
- Darcy Dressell
- Mel Irish
- Andra Lyons
- Sandra Rushlow
- Joshua Veith
- Younes A

About the Author

Frederick Maxwell Banker is a historical and political science writer who graduated from Ferris State University with a bachelor's degree with a major in history and a minor in political science in 2022. Frederick has been writing the book *The South China Sea War: The War No One Should Want* for two years. Frederick enjoys reading books and producing videos for his YouTube channel, @FrederickBanker.

www.ingramcontent.com/pod-product-compliance
Lightning Source LLC
Chambersburg PA
CBHW072008120726
48054CB00004B/12
9781968761561